Energy Management in Industry

Energy demand reduction is fast becoming a business activity for all companies and organisations because it can increase profits regardless of the nature of their core activity.

The International Energy Agency believes that industry could improve its energy efficiency and reduce carbon dioxide emissions by almost one-third, using the best available practices and technologies.

This book looks at the many ways available to energy managers to achieve or even exceed this level of performance, including:

- baselining consumption
- planning a monitoring and verification strategy
- metering (including smart, wireless metering)
- energy supply management
- motors and drives
- compressed air and process controls.

It also looks at topics covered in greater detail in its companion volume, *Energy Management in Buildings*: insulation, lighting, renewable heating, cooling and HVAC systems.

Uniquely, it includes a whole chapter on greening data centres. Further chapters examine minimising water use and how to make the financial case, both to prioritise measures for cost-effectiveness, and to get management on board.

This title is aimed at all professional energy, industry and facilities managers, energy consultants, students, trainees and academics, and can be read alongside training for ISO 50001 – Energy Management Systems. It takes the reader from basic concepts to the latest advanced thinking, with principles applicable anywhere in the world and in any climate.

David Thorpe is community manager of Sustainable Cities Collective, an online community for leaders of major metropolitan areas, urban planning and sustainability professionals. Until 2013 he was for 13 years News Editor of *Energy and Environment Management* magazine, for which website he also wrote a weekly op-ed column. He is also the author of several books and hundreds of articles on related subjects. Formerly director of publications at the Centre for Alternative Technology, he has written two other books in the Earthscan Expert series, *Sustainable Home Refurbishment* and *Solar Technology*, and several B2B ebooks for the publisher DoSustainability. He runs his own sustainable media consultancy, Cyberium, manages the Green Deal Advice website, and blogs regularly as The Low Carbon Kid. Find him on Twitter @DavidKThorpe.

Earthscan Expert Series
Series editor Frank Jackson

Solar:

Grid-Connected Solar Electric Systems
Geoff Stapleton and Susan Neill

Solar Domestic Water Heating
Chris Laughton

Solar Technology
David Thorpe

Home Refurbishment:

Sustainable Home Refurbishment
David Thorpe

Wood Heating:

Wood Pellet Heating Systems
Dilwyn Jenkins

Renewable Power:

Renewable Energy Systems
Dilwyn Jenkins

Energy Management:

Energy Management in Buildings
David Thorpe

Energy Management in Industry
David Thorpe

Energy Management in Industry

The Earthscan Expert Guide

David Thorpe

Series Editor:
Frank Jackson

LONDON AND NEW YORK

First published 2014 by Routledge

2 Park Square, Milton Park, Abingdon, Oxon, OX14 4RN
605 Third Avenue, New York, NY 10017

Routledge is an imprint of the Taylor & Francis Group, an informa business

First issued in paperback 2020

© 2014 David Thorpe

The right of David Thorpe to be identified as author of this work has been
asserted by him in accordance with sections 77 and 78 of the Copyright,
Designs and Patents Act 1988.

All rights reserved. No part of this book may be reprinted or reproduced
or utilised in any form or by any electronic, mechanical, or other means,
now known or hereafter invented, including photocopying and recording,
or in any information storage or retrieval system, without permission in
writing from the publishers.

Disclaimer: Neither the author nor the publisher takes responsibility for
any errors that happen as a result of reading this book. All work on
electrical equipment should only be carried out by appropriately qualified
personnel. Equipment manuals should always be referred to and followed.
The operating parameters of equipment are discussed here only in general
terms.

Trademark notice: Product or corporate names may be trademarks
or registered trademarks, and are used only for identification and
explanation without intent to infringe.

British Library Cataloguing in Publication Data
A catalogue record for this book is available from the British Library

Library of Congress Cataloging in Publication Data
Thorpe, Dave, 1954-
   Energy management in industry : the Earthscan expert guide /
David Thorpe. — First edition.
     pages cm. — (Earthscan expert series)
   Includes bibliographical references and index.
   1. Industries—Energy conservation. 2. Industries—Energy consumption—
Management. I. Title.
   TJ163.3.T46 2014
   658.2′6—dc23
                                    2013021247

ISBN 13: 978-0-415-70647-6 (hbk)
ISBN 13: 978-0-367-78743-1 (pbk)

Typeset in Sabon by Keystroke, Station Road, Codsall, Wolverhampton

Contents

Illustrations

Figures

Tables

Preface

I spent 13 years from 2000 as News Editor of *Energy and Environment Management* magazine, and have worked in the industry for over 20 years. This title, and its companion volume, *Energy Management in Buildings*, distill everything I have learnt during this time, and it is astonishing how much of it is not new, even though there is now more evidence that it works.

The arguments for energy efficiency are always compelling: profits increase as costs reduce. Renewable energy receives lots of headlines, but the most cost-effective carbon savings are achieved through energy efficiency. This is finally dawning for organisations around the world, particularly those which want to reduce their exposure to volatile fuel and electricity prices and reduce their carbon emissions.

This book attempts to address general issues for most branches of industry. It argues that serious energy managers should become accredited with the international standard for energy management, ISO 50001. There are plenty of training courses leading to this qualification. This book is intended to complement them, the standard and similar standards (see Table 0.1).

It is often said that we already have the technology to slash carbon emissions and live sustainably; it's persuading organisational managers that it's worthwhile which is the real challenge. Without this, efforts will fail. I come across report after report, for example, from the Carbon Disclosure Project and from the World Business Council for Sustainable Development, saying that organisations which adopt low or zero carbon and sustainability targets as a core part of their business strategy perform above average in other respects. The world needs to hear these messages.

This book also argues for energy efficiency to be seen in the context of resource efficiency and, ultimately, a closed loop system of production, in a world of dwindling resources, especially water resources. Because investment cycles take 15 years at the least, business is far more scared of climate change than politicians. And business is well aware of resource scarcity. The institutions and businesses that will be around in 20 or 40 years' time will be the ones which have taken this message on board, who concentrate on getting more from less.

So I hope these books, and the series of which they form a part, serve some use. I wish every energy manager, whether fledgeling newbies and students, or jaded veterans picking up new tricks, the best of luck. If there is anything they wish to include in a future edition, please would they get in touch.

I would like to thank Alan Aldridge (Director of the Energy Services and Technology Association, ESTA), Dr John Ryan (Director of Certification Europe), as well as others who have commented on drafts of the book, including especially the series editor Frank Jackson, and Helen Adam, plus the past editors of *Energy and Environment Management* magazine, including Nick Bent, who was Editor until recently.

David Thorpe, Wales, May 2013

Introduction

Energy management: the secret of a thriving organisation

Executives need to learn that acting on sustainability issues such as energy management and water scarcity will increase profits and provide long-term security. Companies which have made investments in this area will uncover more innovative products, processes and business models.

This, together with increasing amounts of legislation, means that more and more energy managers are being required in industry. The International Energy Agency estimates industry could improve its energy efficiency by almost one-third and reduce carbon dioxide emissions by up to 32 per cent by using the best available practices and technologies. Currently, industry accounts for 28 per cent of global energy consumption and 32 per cent of carbon dioxide emissions.

Cost savings from energy management are significant, with paybacks estimated in three years in OECD countries and five years in non-OECD countries. The main reasons that people don't get involved are listed in a recent Carbon Trust report, as follows:

- low priority given to energy efficiency by the company;
- capital constraints;
- short investment payback period requirements;
- lack of expertise in energy efficiency;
- complex internal decision making processes that favour simple, compelling propositions;
- other, including fear of disruption of daily business.

The barriers were greatest for the smaller organisations.[1]

Energy efficiency

Energy efficiency is frequently described as the 'low-hanging fruit'. It is estimated that the global market value of innovative products in this sector could reach around £488 billion by 2050.

In the UK alone, innovative energy-saving measures in non-domestic buildings could save $18MtCO_2$ by 2020 and $86MtCO_2$ by 2050, depending on the rate at which the measures can be deployed.[2] In the USA, the American Energy Manufacturing Technical Corrections Act was passed at the end of 2012, a modification of the Enabling Energy Savings Innovations Act. This promises to produce a boom in the sector. The US market for energy efficiency and services topped $5.1 billion in 2011, according to Pike Research, and is now expected to reach $16 billion in sales by 2020.

Figure 0.1 The context for energy efficiency within other concepts.

Source: Author

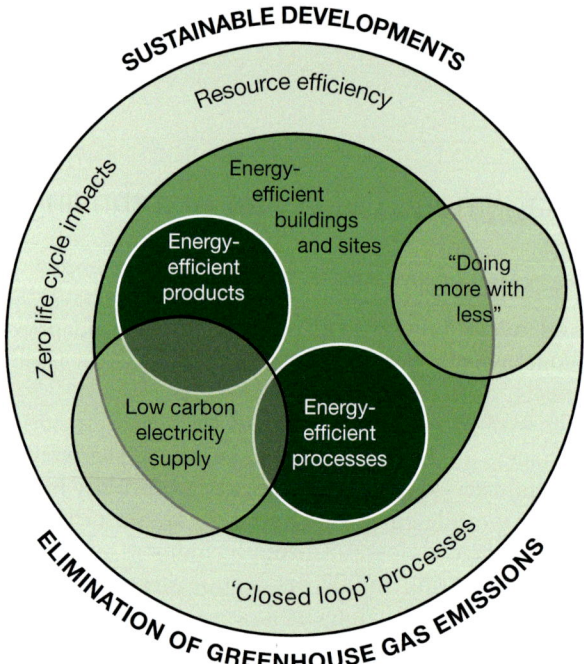

Energy managers are catalysts for this change.

Energy efficiency is a subset of resource efficiency. For example, reusing waste materials saves the purchase of new ones. Redesigning products, or the production line, so that fewer resources are used will have the same effect. Both result in the use of less energy. Similarly, water efficiency not only results in lower water and sewerage bills, but it uses less energy in pumping or heating, producing double wins.

Emissions abatement opportunities in UK industries offer tremendous potential to generate energy, save carbon and reduce the cost of operations. The abatement potential in the key emitting industries in the UK is in the range of 270–500M tonnes with cost savings of £17–32 billion by 2050. The largest single emitter is the chemicals sector.

Based on industry interviews and research, a set of abatement technologies have been identified which principally include alternate process technologies, low carbon substitutes, recovery and recycling, and additional technologies like Carbon Capture and Storage (CCS). Successful innovation in these technologies has the potential to save energy costs by between £4 and £10 billion, especially as prices of conventional fuels rise. Together with savings in carbon abated of between £13 and £22 billion, this could save the UK a total of £17–32 billion by 2050.[3]

Innovation in this area is considered to be critical in order to avoid the risk of industries becoming uncompetitive, which would move production overseas.

This book looks at many general ways of reducing energy in industry, but there are also specific technologies applicable to different industries which this book is not able to go into for reasons of space. These include the following.

Chemicals

- membrane separation for bio-processing;
- alternative process technologies;
- the use of Oxygen Depolarized Cathodes (ODC) in the Chlor-Alkali process, which has the potential to lower energy use by about 30 per cent for membrane cells;
- using bio-based feedstocks (monomers/polymers derived from crops, micro-organisms and fermentation products) and enzyme (biocatalysts) to produce chemical products;
- the use of carbon capture and storage (ammonia and ethylene plants only due to high concentration of CO_2 in flue gas);
- advanced recovery and recycling to reduce demand for virgin polymers and other raw materials – the resulting chemical intermediates from plastic recycling are suitable for use as feed stock for new petrochemicals and plastics.

Iron and steel industry

- improved process controls and neural network-based technologies for existing blast furnaces and electric arc furnaces;
- smelt reduction, which can significantly lower coal use in blast furnaces;
- electrolysis and use of continuous strip production and charging in electric arc furnaces;
- using biomass and natural gas as a partial substitute for coal;
- the use of Top Gas Recycling with Carbon Capture and Storage (TGR with CCS) via chemical absorption or membranes – the use of pure oxygen for combustion rather than ambient air will produce higher concentrations of CO_2 which will be easier to capture;
- reusing structural steel components and recycling steel scrap in electric arc furnaces.

Food and drink industry

- modified food products and processes like low-process animal feeds;
- reduced thermal mass of baking tins;
- water reuse and alternatives to existing heating;
- anaerobic digestion;
- cooling and cleaning process technologies such as UV pasteurisation of milk (non-homogenised milk);
- the use of efficient gas engines for refrigeration and use of ice slurry for cleaning pipes (ice pigging).

Cement

- using fluidised bed kilns, which efficiently combust low-grade coal and increase the heat recovery efficiency between the components;
- replacing clinker with alternatives (furnace slag, fly ash, volcanic rock, limestone, etc.);

Low carbon concrete

A major disadvantage of concrete is its large carbon footprint: one tonne of cement resulting in the emission of approximately one tonne of CO_2. Furthermore, only 50 per cent is currently recycled for use in new building projects (compared to up to 99 per cent for structural steel). Down-cycling does help to reduce the use of aggregates, but does not help to reduce the supply of materials needed for new concrete.

Low carbon concrete is a new product which involves the accelerated carbonation of magnesium silicates under high temperature and pressure, with the resulting carbonates then heated at low temperatures to produce magnesium oxide, with the CO_2 generated being recycled back in the process. The use of magnesium silicates eliminates the CO_2 emissions from raw materials processing. In addition, the low temperatures required allow the use of fuels with low energy content or carbon intensity (i.e. biomass), thus potentially further reducing carbon emissions. Furthermore, production of the carbonates absorbs carbon dioxide by carbonating part of the manufactured magnesium oxide using atmospheric/industrial CO_2. Overall, manufacturers claim that making one tonne of cement using this method absorbs up to 100kg more CO_2 than it emits, making it a carbon-negative product.

Steel recycling

Steel is a robust and long-lasting material, despite being moderately energy intensive, and can be easily purchased as a recycled product. Recycling steel saves energy, CO_2 and resources by displacing the need to make more steel from virgin sources. Over 85 per cent of steel is recycled at the end of its life. By saving remelting, reuse is the most environmentally advantageous approach at the end of a building's life. The energy used in producing steel from recycled steel is approximately one-third of that for new steel.

- the use of low carbon cement produced from magnesium silicates, dolomite or geopolymers (sourced from blast furnace slag and pulverised fly ash);
- the use of Carbon Capture and Storage via chemical absorption or membranes – the use of pure oxygen for combustion rather than ambient air will produce higher concentrations of CO_2 which will be easier to capture.[4]

Legal requirements

There are numerous statutory requirements motivating the saving of energy. The newly published British Energy Efficiency Strategy looks to achieve 196TWh of energy savings in 2020, with a reduction of around 11 per cent over the business-as-usual baseline, and a reduction in carbon emissions of 41MtCO$_2$. The Energy

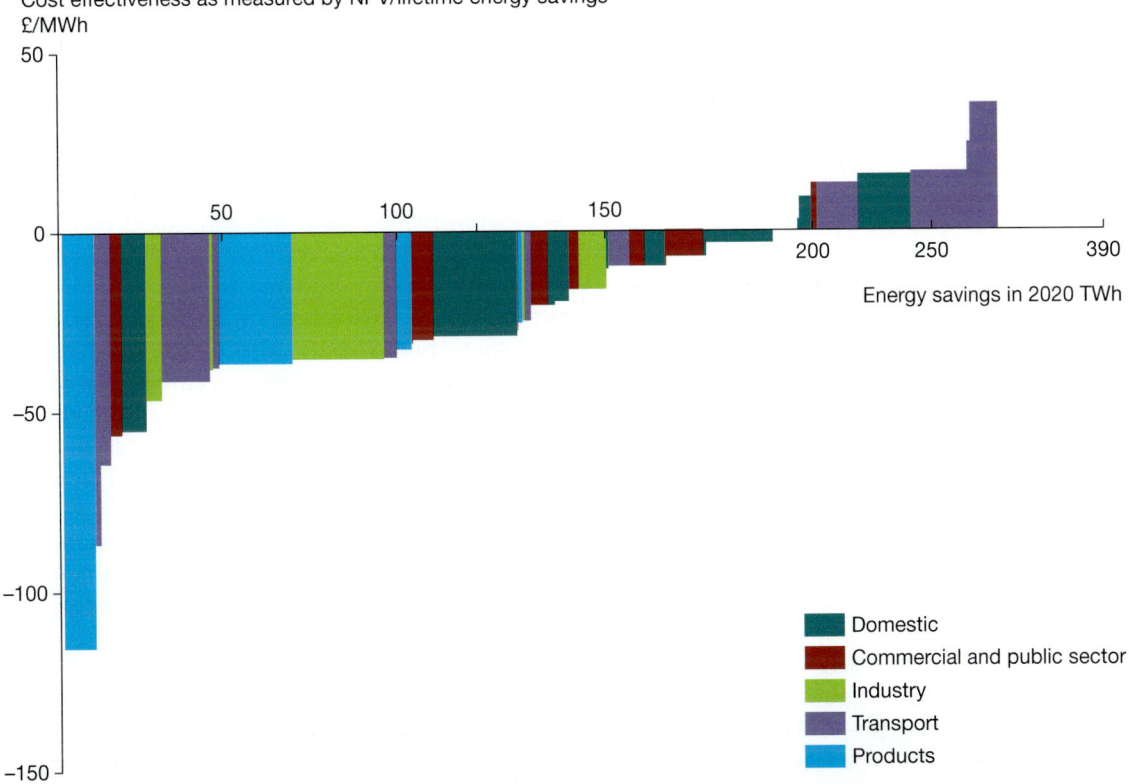

Cost effectiveness as measured by NPV/lifetime energy savings
£/MWh

Energy savings in 2020 TWh

Domestic
Commercial and public sector
Industry
Transport
Products

Figure 0.2 The UK government's 2020 Energy Efficiency Marginal Abatement Cost Curve (EE-MACC). The graph quantifies the lifetime cost benefits of various energy efficiency measures across different sectors, and is discussed in more detail in Chapter 13. The y-axis represents the cost-effectiveness of a measure, each of which is represented by an individual coloured bar. Any measure which costs more than it saves over its lifetime is represented by a bar which goes over the horizontal axis. The overall message is that the vast majority save money over their lifetime. The net present values are calculated in 2012 terms. The EE-MACC is based on an estimate of the feasible roll-out of energy efficiency measures and takes into account supply constraints for energy-efficient products, only including technology that is already available in the market.

Source: UK Energy Effiiciency Strategy, November 2012

Management Alliance, a forum for the UK's energy management companies and industry bodies, foresees a huge growth in the sector.

The EU's Energy Efficiency Directive has a target of 20 per cent energy savings for the EU as a whole by 2020. It requires energy audits and energy management by large firms, and stipulates that 3 per cent of public buildings that are owned and occupied by central government must be renovated every year. The recast EU's Energy Performance of Buildings Directive (EPBD) was transposed into national legislation in 2012 and sets energy performance standards for new buildings and benchmarks for existing buildings. All buildings must have an Energy Performance Certificate (EPC) available when offered for sale or rent.

In the USA Energy Efficiency Resource Standards (EERS) require that utilities or non-utility programme administrators achieve a percentage reduction in energy sales from energy efficiency measures. The Obama Administration has set a goal of doubling the nation's energy productivity by 2030. Manufacturing accounts for 30 per cent of all energy consumed in the United States, and buildings for approximately 40 per cent. The Administration is set to spend about $1.5 million up until 2015 to create new technical assistance partnerships (TAPs) to help manufacturers implement combined heat and power (CHP) programmes in their facilities to take advantage of lost energy.

As far as existing legislation goes, the Enabling Energy Savings Innovations Act (H.R. 4850) waives certain insulation standards placed on some components of walk-in coolers and freezers as set by the Energy Policy and Conservation Act (EPCA) of 1975. But, crucially, it directs the Secretary of Energy to report to Congress on the deployment of industrial energy efficiency within one year of the enactment of the Act (i.e. September 2013), with advice on removing barriers to the deployment of energy-efficient technologies. It also sets Federal energy management and data collection standards, including a web-based tracking system to certify compliance with certain energy and water measures, and requires a study of the perceived economic benefits of providing the industrial sector with Federal energy efficiency matching grants, and estimated energy and emission reductions, especially for water heaters/water-heating technologies. The American Energy Manufacturing Technical Corrections Act of 2012 (H.R. 6582) establishes standards such as best practices for 'smart' electric meters in the Federal government, and sets Federal energy management and data collection standards.

Looking into the future, the Energy Savings and Industrial Competitiveness Act of 2013, if passed, would create energy efficiency financing programmes for commercial buildings and the industrial sector and would require the Federal government to increase efficiency, among other energy conservation measures.

Energy security and procurement

From the point of view of an energy manager, energy security implies securing a constant supply at a realistic price. This is normally achieved via a procurement strategy, but increasingly energy managers are also implementing their own generation schemes, typically from one or more renewable energy or combined heat and power sources. A typical procurement contract will be selected on the basis of competitive tendering. Larger companies typically secure the services of an energy procurement consultant whose responsibility is to protect the organisation from risk. It is important to choose a broker who is competent in reducing the organisation's carbon footprint.

One of the easiest ways for an organisation to slash its carbon footprint is by a green electricity purchasing contract which guarantees a supply of renewable electricity. There are two types of such a contract: in the first, the utility guarantees to purchase an amount of renewable electricity from a supplier equivalent

Typical priorities

1   behaviour change – switch off, turn down;
2   draughtproof – remove leaks;
3   insulate to as high a standard as possible;

> 4 double- or triple-glaze;
> 5 eliminate thermal bridges;
> 6 make as airtight as possible;
> 7 install passive stack ventilation with night cooling or, if not possible, mechanical ventilation with heat recovery;
> 8 supply the remaining energy renewably only where appropriate.

to that supplied. In the second, the electricity supplied actually comes directly from a renewable source such as a wind farm, waste-to-energy plant or landfill gas-fired turbine. The latter is to be preferred, as it provides more of a market stimulus to invest in new renewable energy generation plant.

Standards

This book recommends obtaining the professional qualification associated with energy management, namely ISO 50001, or any of the other standards (see Table

Table 0.1 Standards in energy management internationally

Country	Standard	Description
Worldwide	ISO 50001:2011	The only international framework for industrial plants, commercial facilities or entire organisations to manage energy, including all aspects of procurement and use. Takes account of many of the schemes below.
EU	EN 16001:2009	Covers requirements of energy management systems, now obsolete and replaced by the above.
USA	ANSI/MSE 2000	Voluntary standard for an energy management system. Covers the elements required to ensure continual improvement, sustain savings from energy projects and a strategic energy management plan.
Europe	EN 16247	Defines the attributes of a good-quality energy audit; mandated by the European Commission for Energy Efficiency Directive.
Australasia	AS/NZS 3598-2000	An Energy Audit Standard that represents good practice for energy auditing.
Denmark	DS 2403	Sets out requirements for an energy management system (2008).
Ireland	IS 393:2005	Energy Management Systems Standard to help organisations integrate energy management into their business structures. Shares common management system principles with the Environmental Management System Standard ISO 14001 (superseded).

0.1). It is intended to be read alongside the process of training to attain such certification.

EN standards apply to the whole of Europe. ISO standards are global. A new standard for energy audits, ISO 50002, and a complete set of EN 16247, will be available by early 2014. These represent an improvement on the current standard, EN 16247-1. EN 16247 is broken down into: EN 16247-1 (general); EN 16247-2 (building); EN 16247-3 (process); EN 16247-4 (transport); and EN 16247-5 (qualifications of energy auditor). However, there are differences between EN 16247-1 and ISO 50002. In principle, EN 16247 is focused on energy efficiency, whereas ISO 50002 is aligned to ISO 50001; in other words, it takes the more holistic view of energy performance, defined as energy use, energy consumption and energy efficiency. At some point in the future, Europe will need to decide to withdraw EN 16247-1 and realign EN 16247-2, -3, -4 and -5 to ISO 50002.

Of all the published ISO standards, over 155 relate to energy efficiency and renewables, with many more in development. They cover both generic subjects such as energy management and energy savings, as well as sector-specific solutions for buildings, IT, industrial processes and transport, among others. ISO standards for renewables tackle subjects such as bioenergy, biofuels and solar power. To develop them, ISO, the International Standards Organisation, works closely with key organisations in the energy field, such as the International Energy Agency (IEA), the International Electrotechnical Commission (IEC), the World Energy Council (WEC) and the Efficiency Valuation Organization (EVO), as well as sectoral organisations like the International Commission on Illumination (CIE).

Basic concepts, terms and definitions

The vocabulary of an energy manager revolves around kilowatt-hours, kilograms of carbon dioxide, U-values, square and cubic metres, and units of currency. Basic terms and definitions, with other conversion factors and so on, are given in the Appendix at the back of the book.

The scope of this book

This book takes the reader on the journey from novice to expert in a logical fashion. Chapter 1 covers establishing the baseline and setting the foundations in place for a measurement and verification strategy. Most of the book covers implementation of this strategy, beginning with metering in Chapter 2. Chapter 3 discusses making buildings more energy efficient through insulation and super-insulation. Chapter 4 looks at satisfying lighting needs, first maximising the use of daylight, then energy-efficient lighting and controls. Cooling and heating strategies are examined in Chapter 5, first from the point of view of minimising the need for artificial input and heat reclamation, and then with different solutions for sustainable heating and cooling. Complete heating, ventilation and air-conditioning (HVAC) systems are discussed in Chapter 6, both from the point of view of specification and maintenance. Chapter 7 looks at ways of making the most of incoming energy sources, particularly electricity.

The book then moves on to machinery and technology used widely in industry, including motors, drives, compressed air, etc., (Chapter 8), refrigeration (Chapter

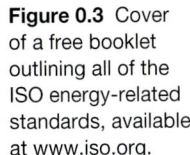

Figure 0.3 Cover of a free booklet outlining all of the ISO energy-related standards, available at www.iso.org.

Source: ISO

ISO&energy

Working for a cleaner, sustainable future

Energy efficiency and renewables are key to meeting the world's energy demand and reducing up to 40 % of carbon emissions by 2050

ISO standards represent consensus on concrete solutions and best practice for energy efficiency and renewables

ISO standards open up markets for innovations that address the energy challenge

9), followed by a look at process controls in Chapter 10 for optimising the use of energy on production lines. Chapter 11 discusses servers and data centres, followed by a look at minimising the use of water through attention to leaks, pumps, efficiency and distribution systems (Chapter 12). Satisfying remaining electricity needs from renewable sources is not covered in this book, but is dealt with in companion volumes, including *Energy Efficiency in Buildings* which has much more on this topic. Finally, the energy manager is given tips on enthusing management about the energy-saving agenda with advice on how to make the financial case (Chapter 13).

Energy-efficient transportation

We will make brief reference to this, since it could come within the remit of an energy manager where a concern operates fleets of lorries, vans or cars and wishes to lower its entire greenhouse gas emissions. Generous tax breaks are often allowed for the purchase of low emission vehicles, and they are cheaper to run, representing a double win. Hybrid or electric vehicles may be favoured, where the electricity is supplied by renewable energy.

The further an electric vehicle travels each day, the more cost-effective they are to their owners, as long as they are frequently charged. A study by the International Transport Forum[5] found that electric passenger cars currently cost €4,000 to €5,000 more to their owners than an equivalent fossil fuel car over the vehicle's lifetime, but because they will travel greater distances, an electric delivery van costs €4,000 less to owners over its lifetime than a similar fossil fuel van.

Forklift trucks represent a particular opportunity for changing to fuel cell-driven vehicles when replacing battery-powered vehicles, owing to the reduced charging time and longer range. Marks & Spencer, Coca-Cola, Walmart and FedEx have converted forklift trucks in their warehouses to fuel cells because they keep going for many times longer than battery-powered trucks. The fuel cell stack manufacturers have designed the fuel cells to be exactly the same size as the battery packs they are replacing, so the trucks can be adapted quickly and simply.

The big challenge

Energy efficiency is interesting because it can be justified in two ways. Even if climate change wasn't a problem, it would be worth saving energy in order to save money. In addition, since climate change is real and one of the biggest challenges facing life on earth, even if it wasn't worth doing to save money it would be worth doing to save carbon emissions. It is a huge challenge facing humanity to reduce and reverse the rate at which greenhouse gases are being emitted into the atmosphere. We hope that this book works as a useful manual or tool to this end for energy managers and students of the subject everywhere.

Notes

1 Exploring the design of policies to increase efficiency of electricity use within the industrial and commercial sectors, Carbon Trust & SPA Future Thinking, Department of Energy and Climate Change, UK, November 2012.
2 UK Energy Efficiency Strategy, Department of Energy and Climate Change, November 2012.
3 The Carbon Plan: Delivering our low carbon future, HM Government, December 2011.
4 Analysing the Opportunities for Abatement in Major Emitting Industrial Sectors, AEA Technology plc, The UK Committee on Climate Change, December 2010.
5 Smart Grids and Electric Vehicles: Made for each other? OECD's International Transport Forum, July 2012: http://bit.ly/N7bMi4.

1

Measuring energy consumption

Energy savings cannot be made on a sustainable basis without appreciation of where energy is being used, for what reason, and what might be influencing consumption patterns. This chapter moves from a basic discussion of how energy is used, through conducting an energy audit, to techniques for regular energy measurement and monitoring. All of this is a prelude to identifying excessive energy use, consequent potential energy conservation measures, and a full measurement and verification (M&V) plan.

Energy units

Energy is described using several different units. The SI unit of energy is the joule. Other units of energy include the kilowatt-hour (kWh) and the British thermal unit (Btu). These are both larger units of energy. One kWh is equivalent to exactly 3.6 million joules, and one Btu is equivalent to about 1,055 joules. Full conversion tables are given in the Appendix at the back of the book, together with tables converting fossil fuel energy use into greenhouse gas emissions and other vital information. These are necessary to convert energy use in different contexts to the same units for calculations.

Energy sources

Energy arrives in a building or facility from many directions: from direct sunshine, from an electricity supply, from fuels such as gas and oil which are burned for heating or to create movement, and from heat arising from activities and processes. Energy is used in many ways: for space heating and cooling, for water heating and process heat, for lighting and equipment such as motors, fans, PCs and chillers, for industrial processes and for transport.

Energy efficiency

Efficiency is measured by the amount of useful work conducted by the energy supplied as it is converted into power. The energy which is wasted in this process appears as heat. This is why LED lighting is so much more efficient: incandescent lighting gives off a lot of heat, which is wasted energy, while LEDs do not.

Similarly, an industrial gas-powered electricity generator will lose 70 per cent of the energy content of the gas as heat unless it is burnt in a combined heat and power generator, which reclaims a further 30 to 60 per cent of the heat.

Energy never disappears when used; it just becomes converted into another form. Eventually, it all becomes heat. It is absorbed into the fabric of the building and equipment, then filters into the environment. It is a general rule that in any system, energy in equals energy out, but before it finally dissipates some of this energy may be reclaimed and reused, or stored in hot water tanks or in batteries, or as hydrogen. Some heat may often be reclaimed from generators and boilers, as we will see in Chapter 5, but there is a natural limit to this process. Sometimes heat generated by equipment is unwanted, as with fridges; an extreme case of this is in data centres, which are discussed in Chapter 11.

The following are the definitions of efficiency for different types of energy use:

- electrical: useful power output per electrical power consumed;
- mechanical: the proportion of one form of mechanical energy (e.g. potential energy of water) that is converted to mechanical energy (work);
- thermal or fuel: the useful heat and/or work output per unit of fuel or energy consumed;

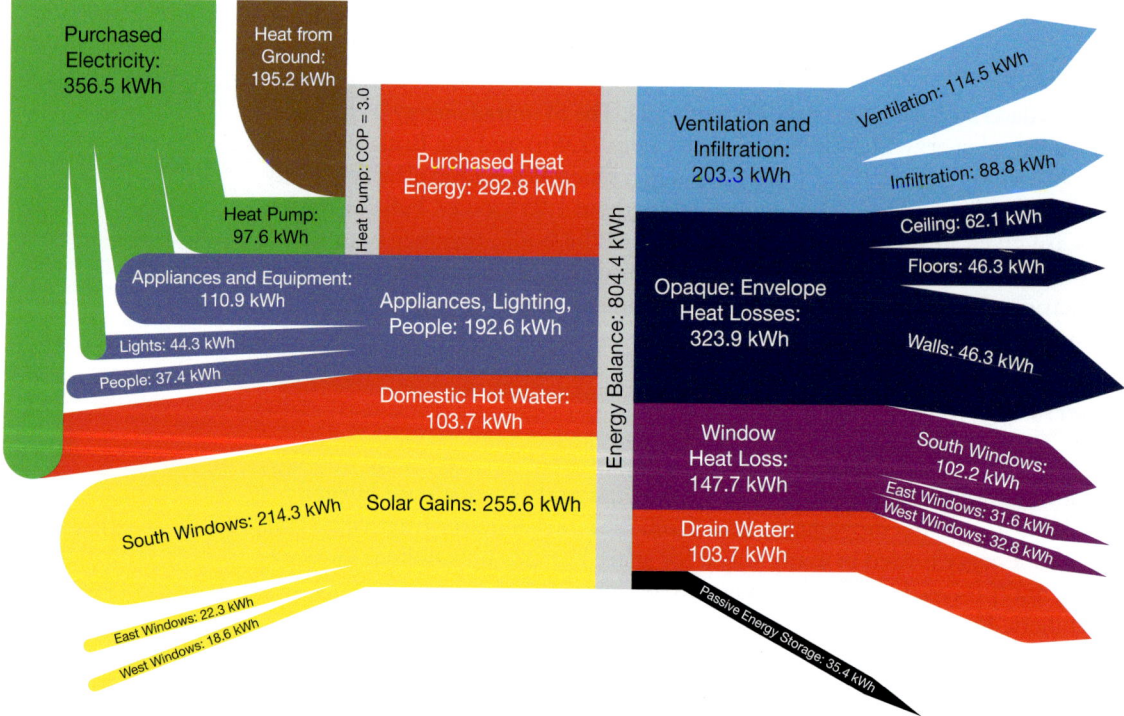

Figure 1.1 Diagam showing the energy balance of a hypothetical building. Sources of energy entering are on the left, and ways in which it leaves are on the right. The total energy balance is quantified in the centre.

Source: 'Preliminary Investigation of the Use of Sankey Diagrams to Enhance Building Performance Simulation-Supported Design', William (Liam) O'Brien of Carleton University, Ottawa.

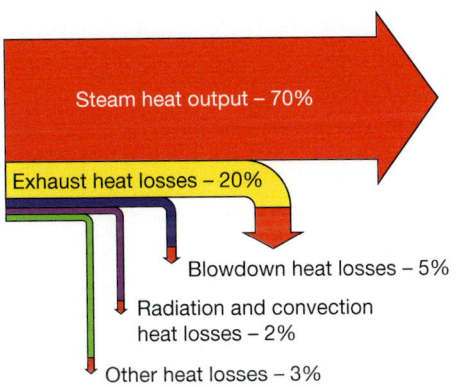

Figure 1.2 Diagram showing the heat losses from a non-condensing boiler. All processes lose energy as heat.

Source: Author

- lighting: the proportion of the emitted electromagnetic radiation usable for human vision;
- total efficiency: the useful electric power and heat output per fuel energy consumed.

It often makes sense to talk about the efficiency of an entire system or process, such as how much electricity is consumed to produce 1,000 widgets. It is extremely important to be clear on the units that are being used and the type of efficiency that is being measured; otherwise confusion may result. For example, there is a difference between Europe and America in the definition of heating value. In the US and elsewhere, the usable energy content of fuel is typically calculated using the higher heating value (HHV), which includes the latent heat for condensing the water vapour emitted by burning the fuel. In Europe, it is the convention to use the lower heating value (LHV) of that fuel, which assumes that the water vapour remains gaseous and is not condensed to liquid water, so releasing its latent heat. Using the LHV, a condensing boiler, which makes use of this latent heat, can achieve a 'heating efficiency' in excess of 100 per cent, whereas, of course, using HHV, this efficiency is around 90 per cent, compared to 70 to 80 per cent for non-condensing boilers.

The aim of energy management is to improve the overall efficiency of the entire systems for which the manager is responsible: to get more work out for the energy put in, and certainly to make use of the free energy that is available, principally from the sun. In calculations to determine building efficiency strategies, when this free energy is called 'passive gains', it needs to be taken into account.

Energy audits

A European and world standard for energy audits, published in October 2012, BS EN 16247-1, explains the process of conducting an energy audit in great detail, defining the attributes of a good-quality energy audit, from clarifying the best approach in terms of scope, aims and thoroughness to ensuring clarity and transparency. It specifies the requirements, common methodology and deliverables for energy audits. It applies to all forms of establishments and organisations, all forms of energy and uses of energy, excluding individual private dwellings, and is appropriate to all organisations regardless of size or industry sector. It was

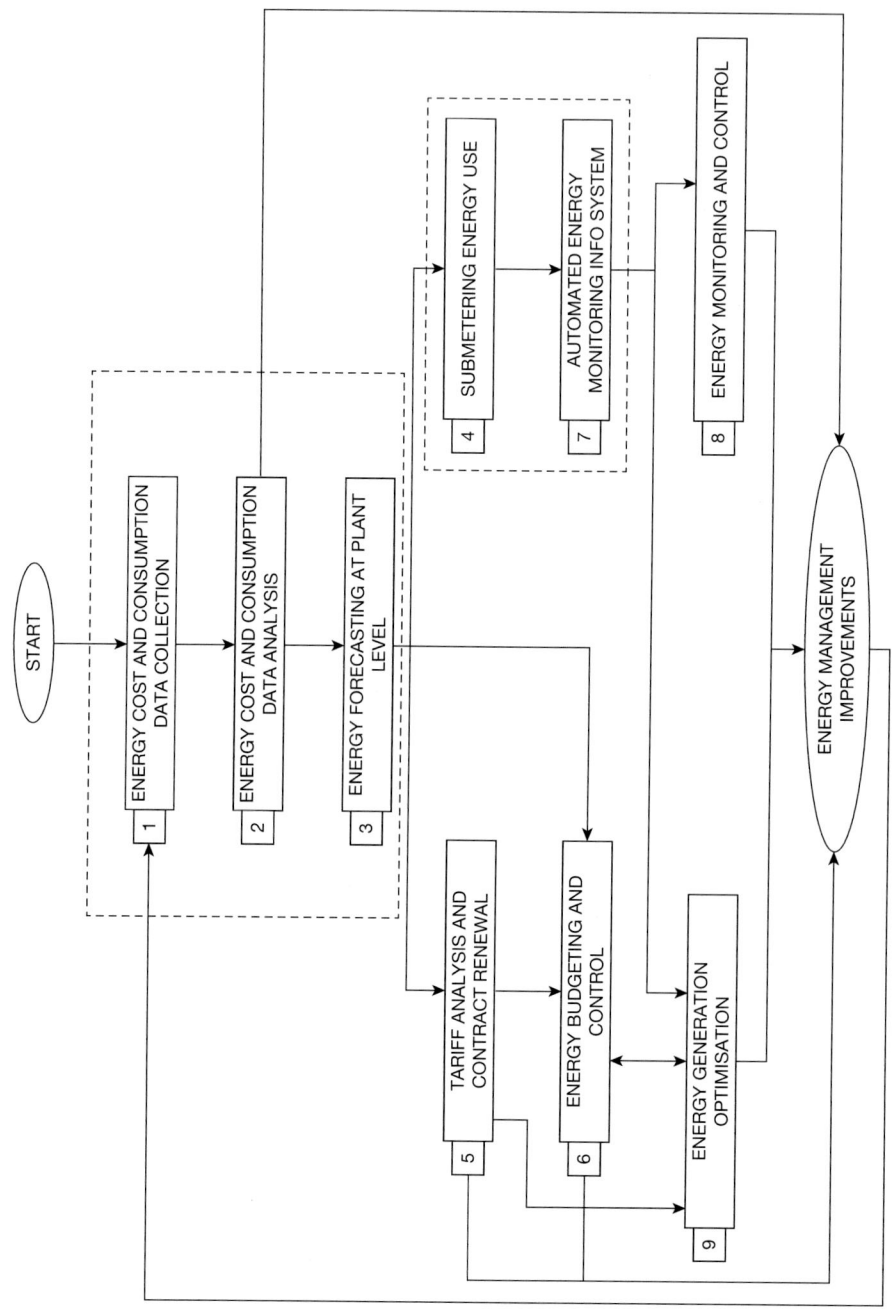

Figure 1.3 The process of auditing, benchmarking and performance optimising.

Source: Carbon Trust

developed by members of ESTA (the Energy Services and Technology Association), the Energy Institute, Institute of Chemical Engineers, and Energy Services and Technology Association in Britain, in response to the 2006 EU Directive and in anticipation of the Energy Efficiency Directive. This mandates member countries to create regular energy audits for large organisations. It complements the energy management system standard ISO 50001:2011 which identifies the need for clear energy auditing. At the time of writing, further, more specific standards are being developed but will be available shortly: energy audits for Buildings (EN 16247-2), Processes (EN 16247-3) and Transport (EN 16247-4).

The first step of an energy audit is to discover where all the energy comes from and where it goes, and to quantify this. It may then be compared to an established benchmark of energy use for the building, company, organisation or facility.

Establishing the baseline

Energy used is either metered or unmetered. Metered information may be gathered from invoices from suppliers and utility companies and from the meters directly. It is necessary to make a list of all meters and sub-meters. This will detail their location, what fuel or energy source they measure, the zone or equipment they serve, the units of measurement, the reading recorded at present (with the time and date), and some other identification such as a serial number. Often, they are photographed. A policy should be set to determine how frequently these are read and by whom. These people need to be taught how to do it, so that everyone is reading the meter in the same way.

Figure 1.4 Modern smart meters provide half-hourly measurements, and even enable the utility companies to provide their customers with cheaper electricity at certain times.

Source: Siemens press picture

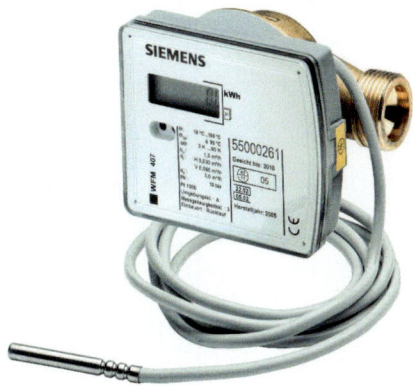

Figure 1.5 Modern heat meters can measure the cumulative heat consumption since starting up the heat meter, cumulative water consumption up until the last due date, and the current return temperature.

Source: Siemens

Fuel that is used for heating and transport will be recorded on invoices. This will include oil, solid fuels, biomass, petrol, etc. Invoices may not be accurate, and it may not be possible to allocate particular purchases to the time period over which that fuel is consumed. Invoices may be duplicates or fuel not delivered but billed for. More accurate information may be obtained from the use of heat meters (e.g. for solar water heating, otherwise unmetered), fuel use measurement, and, in the case of transport, asking drivers to record their fuel purchases and use, mileage, etc.

Having made an assessment of the sources of information for an audit, the next step is to decide the best periods over which to record that information: monthly, weekly, daily or hourly. This will vary from installation to installation. The ideal is that the reporting periods are the same for each type of fuel and power usage. This facilitates and optimises subsequent calculations and the presentation of data in graphical form. Most stakeholders will want to see annual information at the very least, and preferably quarterly and monthly figures. Company directors may wish presentational information to coincide with the company's financial year for use in annual reports.

For example, in the case of fuel use, measurements of stock levels should be taken at the beginning of each week, if that is also the period over which electricity use is to be measured. The amount of fuel use is then the starting amount, plus any amount delivered during that period, less the finishing amount. From the point of view of analysis, weekly, daily, hourly and, preferably, half-hourly information about heating and power requirements is needed, especially for comparing one year with another and one period with another, since months contain an odd number of weeks.

Summary of sources of fuel information

- Utility bills:

  - gas
  - electricity.

- Meter readings:

  - gas
  - electricity
  - heat
  - fuel.

- Transport:

  - driver records and receipts, mileage.

- Fuel:

  - invoices
  - stock level measurements (oil, LPG, wood pellets, coal, etc.).

This information needs to be recorded for each process, zone, building, piece of equipment or vehicle that is being measured. The same recording period (weekly, daily, hourly) needs to be used for each data set. All records are entered into a database. Units need to be converted so that they are compatible (see the Appendix). An additional column may be created in the database to convert these figures into tonnes of carbon dioxide-equivalent emissions. A final column, for a building, would include the floor area in square metres or square feet. This would enable the total energy use to be divided by the total area to give a figure per unit of area. This would enable it to be compared to other similar buildings. The spreadsheet also enables many other calculations and summary reports to be made, and graphs and trends to be seen.

Other information that may be recorded includes the following:

- building occupancy times, and the times of equipment activity;
- process ongoing;
- outside temperature;
- light levels;
- degree days.

Recording the outside air temperature and light levels permits the correlation of energy use with outside conditions and will enable predictions of future energy use based on time of year and weather conditions. Monitoring of interior light levels may also be installed and linked to lighting systems to enable the provision of the optimum level of lighting depending on ambient conditions. This will be considered in more detail in Chapter 4.

Remote fuel management

Estimating the amount of fuel used by fleets, especially from multiple depots, can be a problem. Often it is hard to make sure that the volume used matches up with an invoice, or to be confident of the time period over which the fuel is used, or even the time of delivery. Hydrostatic-level sensors are now available that can be easily dropped into a tank using existing openings, and provide highly accurate volume reading. They are powered using solar cells. They can transmit the volume of fuel in a tank as frequently as every ten minutes, using mobile phone networks and wire-free GSM/GPRS telemetry. This provides a simple way to remotely monitor the levels in fuel storage tanks. The data supplied may be used in conjunction with software which not only monitors and aggregates fuel use across a company or organisation, but can also link to financial packages, provide audits and monitor thefts.

Using the information

The main metric for evaluating energy use tends to be the amount of energy used per unit of output. However, this is not a simple relationship. Just as a business has overheads (rent, fuel, wages, etc.) over and above project-specific costs, so a plant or a site has 'overhead' energy costs regardless of whether a manufacturing process is operating. In other words, if all the production machinery were

Figure 1.6 Production plotted against energy use to determine the baseline and discover discrepancies and opportunities for finding efficiency savings.

Source: Author

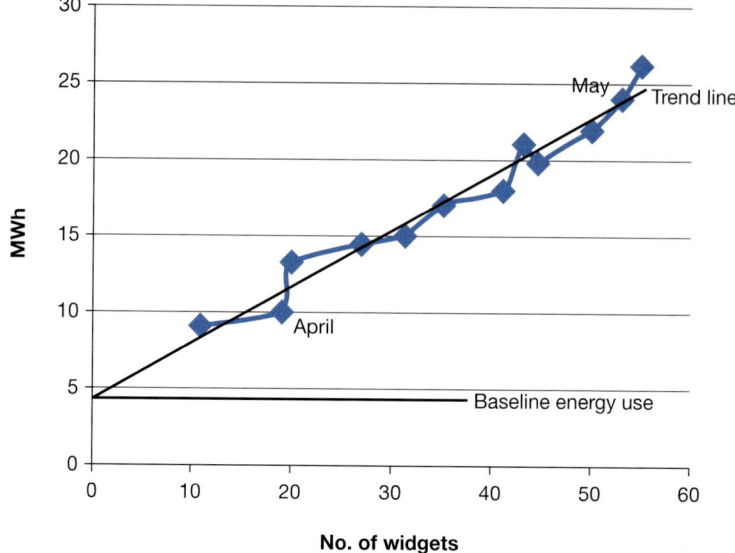

switched off for a month, there would still be energy used; for example, for heating or lighting, computers, etc.

The energy manager primarily uses the gathered energy data to plot a graph of energy use versus production output for each period, as in Figure 1.6, which shows widgets produced against megawatt-hours. The baseline, derived from extending the trend line to the y-axis, represents the minimum amount of energy used: 4MW. In April it looks as though energy efficiency was improved because the number of widgets produced compared to the amount of energy used is below the trend line; whereas in May it is above the trend line and therefore energy efficiency is poorer. But this would not be adduced from a simple calculation of the number of widgets produced per megawatt-hour:

	Number of widgets	MWh	Widgets/MWh
April	19	10	1.90
May	53	24	2.21
Mean			1.98

(The mean figures are taken from all of the readings and are there for comparison.) This makes it look as though May's figures were better than April's. This is because the baseline energy use is not taken into account. Once this figure of 4MW is deducted, the true picture emerges and we can see that April's figures were actually better:

	Number of widgets	MWh	Widgets/MWh
April	19	6	3.17
May	53	20	2.65
Mean			2.65

This is the figure the energy manager should be looking for. He or she can then look for ways to reduce the baseline (overhead) energy use, and discover why the number of widgets produced per megawatt-hour varies.

Independent rating systems

Third-party audits can be helpful in giving confidence and independent verification. In the UK, rating of the energy performance of many public and commercial buildings is compulsory and conducted by independently certified assessors. Depending on the type of installation, they will be Display Energy Certificates or Energy Performance Certificates. The assessment results in the issue of certification as described in figures 1.7 and 1.8. DECs must be publicly displayed and available on a website so that building managers may compare the performance of their building against others. Clients also receive a summary of measures which may be taken to improve the rating.

In the USA, there is no compulsory system for rating buildings or industrial sites. However, the Environmental Protection Agency's ENERGY STAR programme has developed two rating systems for several commercial and institutional building types and manufacturing facilities: the ENERGY STAR Portfolio Manager/Benchmarking Program and the ENERGY STAR, Five-Stage Approach Building Program, which is essentially a measurement and verification (M&V) plan, although it culminates in a certificate.

The Portfolio Manager rating is calculated based on the information a client enters about their facility, such as its size, location, number of occupants, equipment, etc. The system estimates how much energy the site would use if it were the best performing, the worst performing, and every level in between. The system then compares the actual energy data entered to the estimate to determine where the site ranks relative to its peers. These ratings, on a scale of 1 to 100, help to benchmark the energy efficiency of specific buildings and industrial plants against the energy performance of similar facilities; a rating of 50 indicates average energy performance, while a rating of 75 or higher indicates top performance. Those with benchmark scores of 75 or higher are eligible for the ENERGY STAR label for sites, which can be displayed to convey performance excellence to tenants, customers and other users. The ratings are used by energy managers to evaluate the energy performance of existing buildings and industrial and manufacturing facilities. All of the calculations are based on source (primary) energy. The use of source energy is the most equitable way to compare energy performance, and also correlates best with environmental impact and energy cost.

The ENERGY STAR Buildings Five-Stage Approach for energy efficiency looks at a whole building's systems to attempt to achieve average energy savings of 30 per cent for the whole building. The five stages are as follows:

1   Green Lights: Installing energy-efficient lighting systems and controls that can provide substantial energy savings at low cost.
2   Building Tune-Up: Performing a comprehensive tune-up of the entire facility to get it into peak condition.
3   Other Load Reductions: Finding other opportunities for increasing a building's energy efficiency such as purchasing ENERGY STAR office equipment,

installing window film, and adding insulation or a reflective coating to the roof.

4   Fan System Upgrades: Right-sizing fan systems, adding variable-speed drives, or converting to a variable-air-volume system, if appropriate.

5   Heating and Cooling System Upgrades: Replacing chlorofluorocarbon chillers with small, more energy-efficient models to meet the building's reduced cooling loads and upgrading other central plant systems.

The US Green Building Council's LEED (Leadership in Energy and Environmental Design) certification system also provides third-party verification of new and existing buildings. All categories of buildings are suitable for LEED: commercial, institutional, high-rise residential and industrial. Its distinguishing mark is that it addresses the entire life cycle. To earn LEED certification, a project must earn 40 or more points on a 110-point scale. Various rating systems exist to address different types of projects, including health-care facilities, commercial, retail, schools and entire neighbourhoods. Credits are given for the following:

- ecological siting;
- water efficiency;
- energy performance;
- resource efficiency;
- indoor environmental quality.

Measurement and Verification (M&V)

In this chapter we have essentially been introducing the topic of what is known as measurement and verification (M&V). M&V is a set of procedures and methods that identify sources of variability in energy consumption in order to reliably determine energy use and consumption, and compare it against a baseline for a defined system of boundaries. It then becomes possible to construct an M&V plan to reduce consumption.

M&V allows, for example, facility owners, Energy Services Companies (ESCOs) and institutions financing energy efficiency projects to quantify and verify the energy savings performance of improvement measures. They yield a periodic savings figure showing how much less energy has been used than would have been the case if the energy efficiency technology or project had not been implemented. Reported M&V results should also include information on their repeatability, accuracy and reliability. Depending on what is being monitored, system boundaries for reporting might encompass a region, organisation, facility, system, process or list of equipment.

It is extremely useful to use an objective standard for this purpose in order to get external verification of results that will maintain the confidence of management in energy efficiency measures. The International Performance Measurement & Verification Protocol (IPMVP) from the Energy Evaluation Organization (EVO), last updated in 2012, is probably the most widely used system in the world. IPMVP considers a level of expenditure on M&V of no more than 10 per cent of the expected financial saving to be an acceptable maximum level; most projects will be much less than this. In the UK at the end of 2012, the Department of Energy and Climate Change announced that IPMVP would be used as part of the UK's Energy Efficiency Strategy and under the RE:FIT programme for energy efficiency which is being rolled out nationally. It has now been extended to include water savings as well as energy savings. If an energy manager needs to obtain confidence in using IPMVP, the Association of Energy Engineers (AEE) offers the Certified Measurement and Verification Professional (CMVP) accreditation qualification. Given the level of complexity involved in achieving a reliable figure from IPMVP, obtaining this qualification is highly recommended.

Even then, IPMVP may be considered to be rather broad in scope and lacking in detail in some areas, so others may be adopted as well. In the United States, there are three other M&V standards:

1   The ASHRAE Guideline 14-2002: Measurement of Energy and Demand Savings, from the American Society of Heating, Refrigeration, and Air-Conditioning Engineers, which has not been updated since 2002.

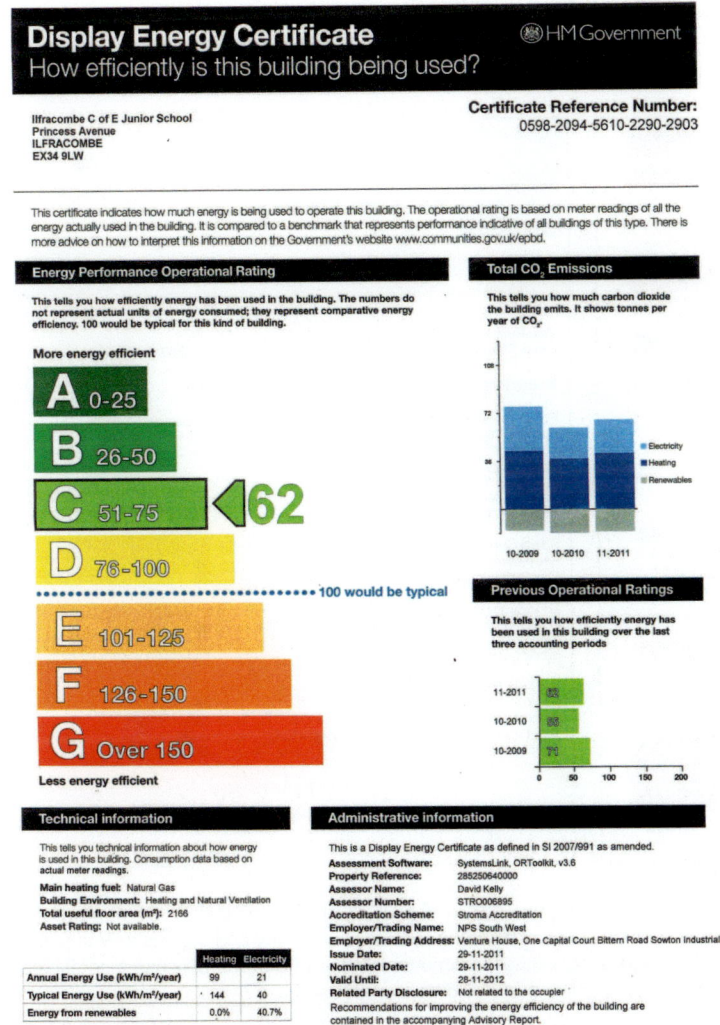

Figure 1.7 Display Energy Certificates are compulsory for all public buildings in England and Wales with a total useful floor area >1,000m² which are frequently visited by large numbers of persons, such as schools and hospitals. Some privately owned buildings that provide services from public funds such as leisure centres, museums and theatres may also need to display a DEC.

Source: Ilfracombe Church of England Junior School, Devon, England

2   The ASME Performance Test Codes from the American Society of Mechanical Engineers, which have a variety of applications for different purposes, and which have various last release dates.

3   The Measurement and Verification for Federal Energy Projects Guidelines from the US Department of Energy, Federal Energy Management Program, last updated in 2008.

Regarding energy efficiency in general, the following published standards exist:

- Energy Management Systems – EN ISO 50001:2011
- Energy Efficiency Services – EN 15900:2010
- Terminology – CEN/CLC TR 16103:2010
- EN 16247-1:2012 Energy audits – Part 1: General requirements
- EN 16212:2012 Energy Efficiency and Savings Calculation, Top-down and Bottom-up Methods
- EN 16231:2012 Energy efficiency benchmarking methodology
- prEN 16247-2 Energy audits – Part 2: Buildings

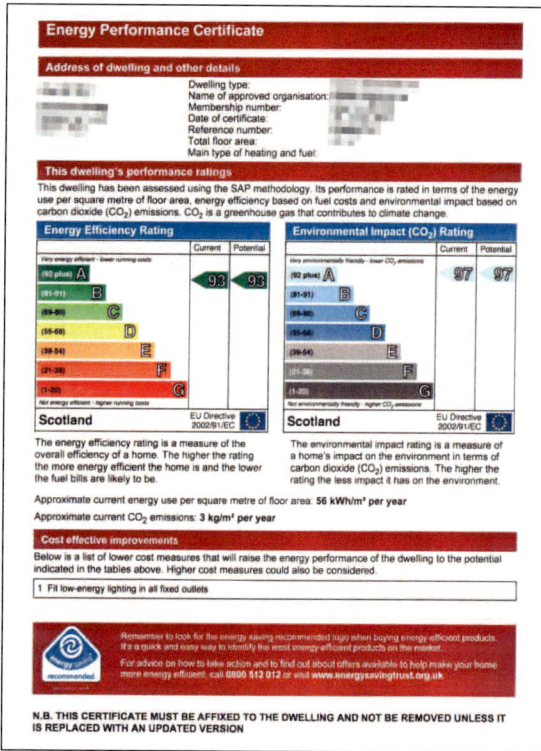

Figure 1.8 Energy Performance Certificates are compulsory for all non-public buildings in the UK. These may contain useful information, but they are just the start of a diligent process of energy management.

Source: Author

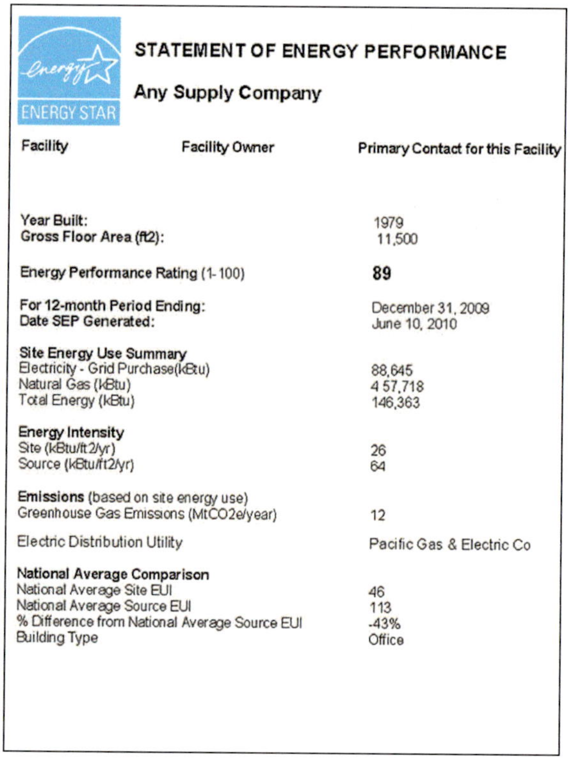

Figure 1.9 In the USA, the EPA ENERGY STAR Portfolio Manager rating system is voluntary and applies only to some buildings in America. At the end of the process a certificate like this would be awarded.

Source: EPA

- prEN 16247-3 Energy audits – Part 3: Processes
- prEN 16247-4 Energy audits – Part 4: Transport
- prISO 50002 Energy audits (UK)
- FprEN 16325 Guarantees

Other countries, including France, the Netherlands, China, Korea, Brazil, Canada and South Africa, are developing their own localised versions of the above standards.

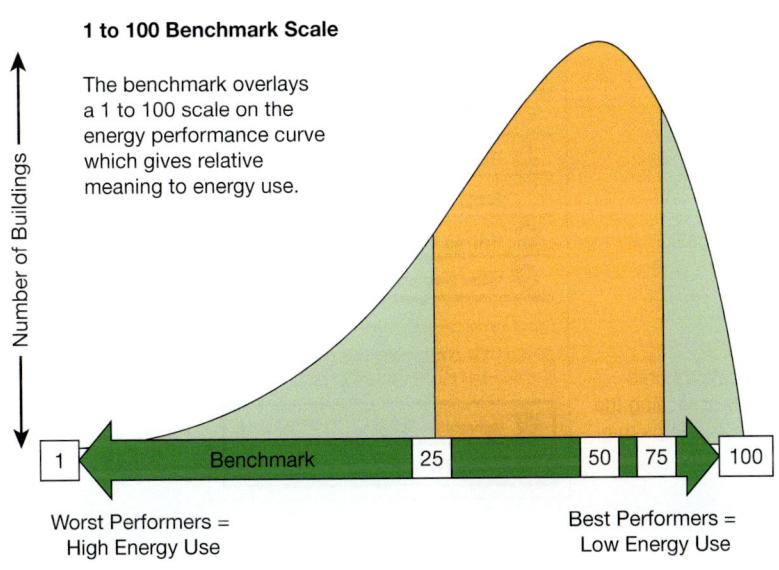

1 to 100 Benchmark Scale

The benchmark overlays a 1 to 100 scale on the energy performance curve which gives relative meaning to energy use.

Number of Buildings

1   Benchmark   25   50   75   100

Worst Performers = High Energy Use

Best Performers = Low Energy Use

Figure 1.10 This curve shows the skewing towards the positive of the proportion of buildings which receive different ratings under the EPA ENERGY STAR Portfolio Manager system for buildings.

Source: EPA

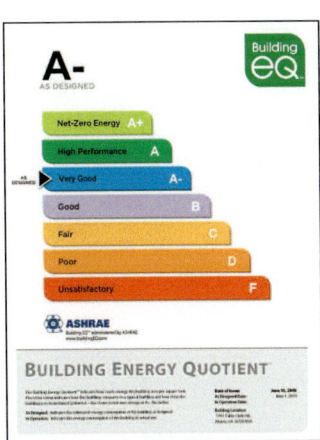

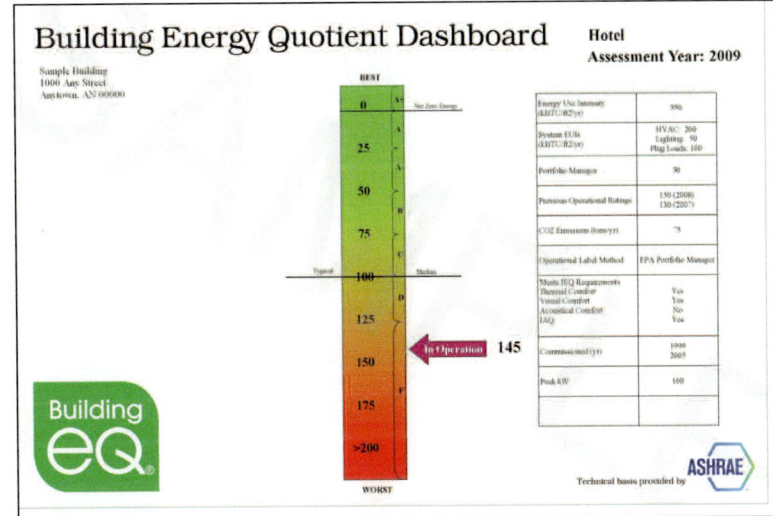

Figures 1.11a and b ASHRAE's building energy quotient label and dashboard, awarded to buildings that volunteer to go through this more stringent, independently assessed American rating system.

Source: ASHRAE

So far in this chapter we have just begun to discuss energy audits. An energy audit with M&V establishes the baseline, and benchmarking compares this with other similar buildings. This will help to develop a strategy for the ongoing recording of energy usage information. To acomplish this, a metering strategy is needed. This is the topic of the next chapter.

Figure 1.12 The type of certificate to be expected from the LEED system of independently assessing the sustainability of new buildings. Existing buildings may also be assessed.

Source: LEED

2
Metering

Energy management is impossible without energy metering. Metering has undergone a revolution in recent years. This chapter looks at the latest technologies, which are increasingly wireless and cloud-based. The first step is an energy audit, which establishes a baseline, and benchmarking compares this with other similar sites. The results will help to develop a strategy for the ongoing recording of energy usage information and an M&V plan.

Electricity and, possibly, heating and fuel use will be recorded on a half-hourly basis. Most metering hardware nowadays is supplied with software that enables interrogation of the data and presentation in visual, graphical form of the actual load profiles over 24-hour, weekly and other periods. The energy manager's eagle eye will be looking for unusual and inappropriate uses of energy, for example, to correlate lighting levels and the heating or air-conditioning profile with the actual occupancy of premises. If a space is being lit or heated when unoccupied, action may be taken to eliminate this. If half-hourly meters are not present, then there is a strong case for having them fitted, because only through this route can detailed analysis and savings be made. An alternative is the use of data loggers (see below).

In large installations, for more detailed analysis, or in concerns with multiple buildings, sub-metering is employed. This involves installing additional meters in places where it is considered beneficial, such as, and depending on the organisation, the chiller room, the data centre, the factory floor, the office suite, or wherever there is a risk of excessive energy being consumed. This a subjective evaluation based on the 'risk of acceptable undetected loss'. For example, a circuit might contain a transformer that loses a proportion of the transmitted electricity in downgrading or upgrading the voltage. Whether this loss is considered 'excessive' and worthy of metering, might depend on the total amount of electricity being consumed.

Meter types

The following is a non-exhaustive list of possible metering solutions that are available.

Electricity

- Regulator-approved meters;
- Measurement Instrument Directive (MID)-compliant meters. MID is a European standard that covers many types of measuring instruments. It is important to check whether a meter is compliant with the regulator (in the UK, Ofgem) or MID or both, for the purpose for which it is intended;
- Combined meters; multifunction microprocessor-based units which measure active and reactive energy in both directions of energy flow, and display instantaneous values such as current, voltage and power.

Electricity meter output options include the following:

- Pulse output: a relay inside the meter that closes momentarily when the meter advances an increment, recorded by metering data loggers.
- MODBUS: here, the data logger asks the meter for its current reading and the meter then responds. MODBUS is the most common protocol in use, but there are others such as M-BUS and BACNET. If no response is received, the data logger can raise an alarm, whereas if a pulse signal fails this could be seen as meaning zero consumption. Pulse-based systems may only log kWh or kVArh (kiloVolt-Amps reactive per hour: the difference between working power (measured in kW) and the total power consumed (measured in kVA)), whereas MODBUS-connected systems can log most parameters that a multifunction meter would measure, such as current, voltage, power factor, frequency, instantaneous power as well as the energy register displays.
- Analogue outputs and alarms: can be programmed to give a 4–20mA signal or similar in sympathy with the amount of instantaneous power being consumed. Others have alarm outputs which can be programmed to signal various conditions.
- Current transformers: used when the primary circuit rating of the feeder that is being monitored is too large to be connected directly to a meter. They are available with primary ratings from 1–5,000A. They can be solid or split core, three-phase low voltage or summation.

Water meters

There are hot (up to 90°C, 190°F) and cold (up to 30°C, 85°F) water meters, which both come in various types: single jet, multi-jet, ultrasonic, screwed connection and Woltmann Flange fitted. Both may be read remotely by connecting up the output contact to an energy management system or building management system.

Heat meters

A heat meter measures the amount of heat (in Btu or kWh) that has been dissipated in a loop of pipework, for example, feeding a heating or processing system. It consists of the following:

- the meter, which measures the flow of liquid in the pipework and has a pulse output connected to the heat integrator or calculator;

- two thermocouples, one in the flow leg of the pipework measuring the hot water entering the loop, the other in the return leg, measuring the temperature of the water after it has been through the loop;
- a heat integrator, to calculate the amount of kWh consumed, using the flow and the temperature difference calculated.

The meter contains an electronic display and an optional pulse output which may be connected to an energy management system. It is possible to obtain heat meters with either MODBUS or M-BUS communications to enable intelligent connections to remote systems. Ultrasonic heat meters remove the need to physically adapt the pipework, enabling much shorter installation times. They incorporate flow and heat calculations in one unit, which saves having to commission two separate meters.

Meter data loggers

These are the core of any energy management system. They collate data from a number of meters (water, electricity, gas or heat meters, etc.) in real time, storing it in local memory. They group these consumption values in time periods, the most common being half-hourly. When viewed as a graph, these values form a profile of the energy consumption for that meter or feeder.

They can take data in either the form of basic metering pulses or intelligently via MODBUS on an RS485 network. Some include basic reports that may be accessed via the internet, using secure File Transfer Protocol (FTP) to a browser, intranet or customisable e-mail, transmitted via a local ethernet network or wirelessly. It is possible to access data within the data logger easily and to interpret the values in a third-party energy management software package or in Microsoft Excel.

Wireless meters

Traditional meters use cabling and consume power. Wireless energy monitoring systems use the internet, and provide two-way communication. They are less intrusive, far easier to fit, and so can be around 50 per cent cheaper to install than cabled meters. This means that every building or unit can have a programmer, as well as sensors for temperature, light and human presence. In this way, wireless control of lights, heating and air conditioning, shading, motors, valves and most technologies becomes possible. This solution can achieve 15 to 30 per cent savings on energy use, and therefore have, on average, a payback period of under 18 months. The latest generation can work with or without batteries, and is largely service free. Systems are modular, allowing for progressive expansion as budgets allow.

The range of wireless subsystems is up to 300 metres outside and up to 30 metres inside buildings. They communicate in the 868Mhz or 315Mhz frequency band at a rate of 125kb per second. Technical means are used to minimise the possibility of transmission errors. Data security protection is supplied as well as protection against attacks.

Figure 2.1 No batteries or wires: this sensor attached to an electric motor relays information about its status and takes its power from the vibrations of the motor itself.

Source: Siemens

Figure 2.2 A mains-powered intelligent wireless input/output module for a wireless energy management system.

Source: WEMS International

Figure 2.3 The interface for an embedded computer system that hosts building energy management application software. For communication with modules, a USB wireless interface module is attached.

Source: WEMS International

Energy harvesting meters

With battery-free systems, energy is harvested from anywhere in the surroundings: linear motion, light, differences in temperature and heat. For example, they can use thermistors to harvest the heat, PV cells to run off light, or piezoelectric crystals to run off pressure changes, such as from vibrations, to send a unique radio message to the receiver module. This modest amount of energy is sufficient to send the wireless signal or turn on a light.

Building energy management systems

From such a metering system it is a simple step to a building energy management system (BEMS: WEMS = wireless energy management system). This is a computer-controlled system that may be used to monitor and control a site's power systems, including lighting, heating, ventilating and air conditioning. Its basic functions include monitoring, controlling and optimising energy use with the aim of matching the use of energy with its application so as to optimise both.

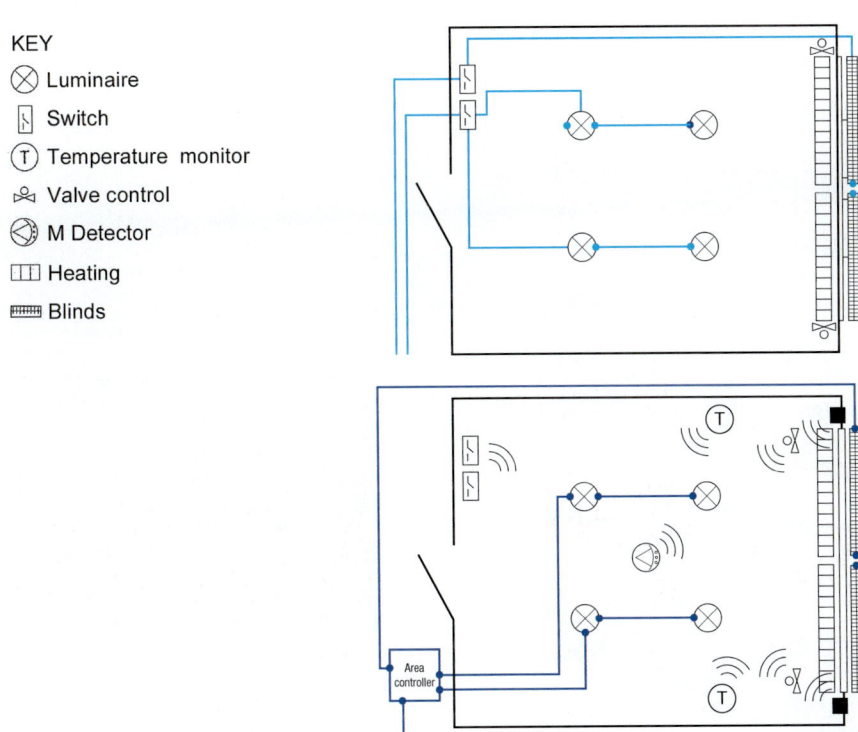

Figure 2.4 A simple wireless energy management system layout for a room. The top diagram shows conventional wiring linking up blinds, valves and heating, lighting and switches. The bottom diagram shows the wireless conversion, with a motion sensor in the middle of the lighting, and two temperature monitors controlling the heating system. The switches are also wireless. Everything is connected to the area controller which also monitors and records activity. A 30 per cent energy saving is claimed, with 70 per cent less cable required.

Source: Author, with thanks to EnOcean

Figure 2.5 A typical schematic diagram for a wireless energy monitoring and control system for a small factory. Readings are collected from sub-meters all around the site and are available to view on screens in the reception area, or anywhere else that is specified.

Source: Author, with thanks to Elcomponent

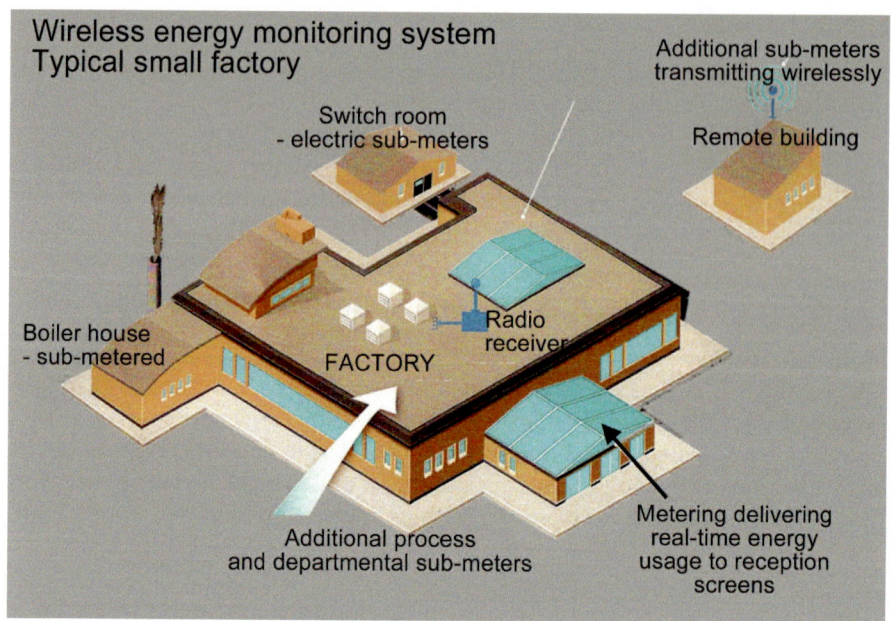

Wireless energy monitoring system
Typical small factory

Additional sub-meters transmitting wirelessly

Switch room
- electric sub-meters

Remote building

Boiler house
- sub-metered

Radio receiver

FACTORY

Additional process
and departmental sub-meters

Metering delivering
real-time energy
usage to reception
screens

Figure 2.6 Splitting a site into zones can help target energy use. This is accomplished using software like this, which can be programmed to automatically take into account occupancy levels, outside temperature and humidity.

Source: Trend Control Systems

This could involve distributing and managing the level of lighting, heating and ventilation to suit the changing levels of activity within different areas of the same building throughout the day, week and year. The power used by process equipment can also be managed.

The level of control may be refined to whatever specification is required, for example, by splitting areas into different zones within, say, an open-plan office, warehouse, large site or factory. Systems are also available that extend the management of resources from energy only to include water, air and steam. So-called WAGES (Water, Air, Gas, Electrical, Steam) systems are intended for industrial applications. Water is a costly resource that can also consume energy, for example, in pumps and heating. We will look at water management in more detail in Chapter 12.

The supply of sophisticated software for BEMS and measurement and verification (M&V) is developing fast, forming a revolution in the tools available to energy managers. Increasingly, software can be cloud-based and involve modular, downloadable apps for many different purposes. For example, apps are available that keep track of tenants' utilities, monitor buildings' energy performance, benchmark energy consumption, compare usage among other buildings

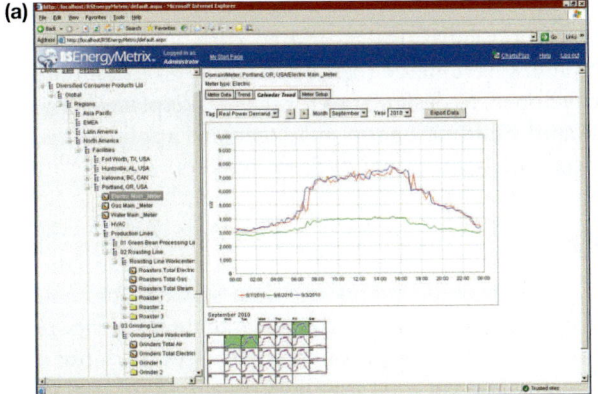

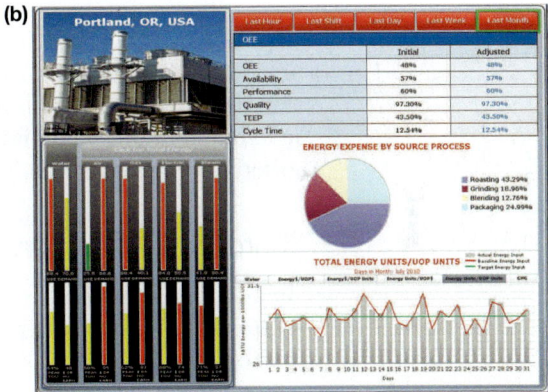

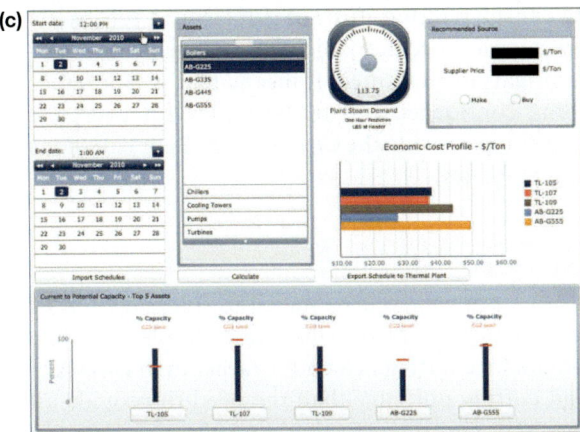

Figures 2.7a, b and c Three stages of use for WAGES (Water, Air, Gas, Electrical, Steam) energy data; from a to c: awareness, efficiency and optimisation.

Source: Rockwell

Figure 2.8 Once the baseline is known, specialist software can allow the exploration of scenarios that determine various energy and cost savings depending on the measures taken.

Source: BuildingIQ

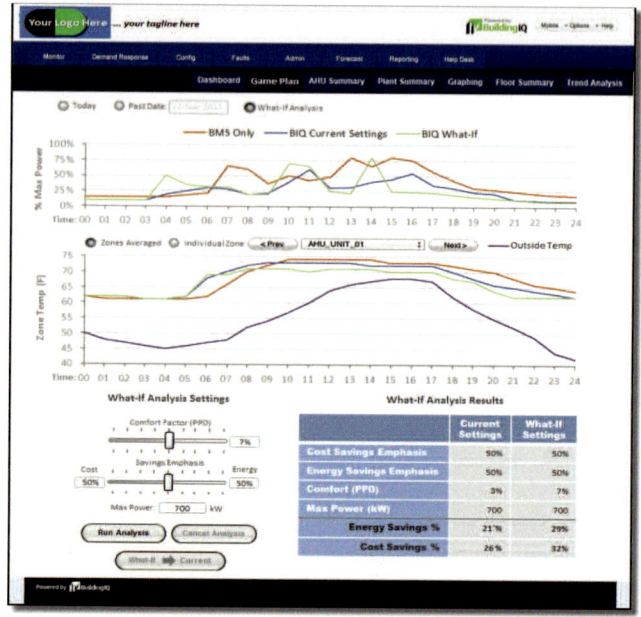

in a portfolio, and display the carbon emissions associated with a building, or group of buildings. Some companies have made their application programming interfaces (APIs) available to developers to design apps for the marketplace. Some combine building intelligence and building energy management applications, offering insight for both financial and operational decision-makers.

Third-party data services

In addition, many utility companies provide energy use data services for their customers. This provides access to a dashboard that enables them to compare performance with baselines and to prepare reports. Such services are now a standard part of contractual arrangements with suppliers, but they do not all provide the same level of quality service, so it may be worth shopping around. On the other hand, it might be considered that it is not sufficient to rely solely on information from the energy supplier because objective, independent monitoring provides a level of risk mitigation guarantee. It means that performance analysis can be verified, yielding confidence and transparency for everyone concerned. In all cases, software should provide the carbon-equivalent emissions for the energy used for reporting purposes.

Anomaly detection

The savings won in the case studies are achieved by judicious interrogation of the data provided by these energy management systems. For example, one may notice that lighting or heating systems are cutting in an hour earlier than necessary at certain times of the year, and cutting out later than the need to provide an acceptable level of comfort. One may also discover that patterns of energy use

Case study: BT

BT (British Telecom) has installed a wireless building energy management system across 2,000 telephone exchanges. These range in size from 400m² to 17,000m². BT anticipates saving 22 per cent of its energy use as a result of this measure. Richard Tarboton, BT's Director of Energy and Carbon, says he believes that this initiative 'has underpinned the success of our smart energy programme at BT. It gives us remote control over the air-handling systems and enables us to optimise energy usage. As a result of the proven success, we have now agreed a contract to expand the programme further.'

Case study: Farmfoods, UK

Farmfoods is a company that specialises in frozen food; therefore, unsurprisingly, freezers are responsible for most of its electricity consumption at its 300 sites. Following the introduction of a wireless building energy management system, it managed to reduce this consumption by 18 per cent. The return on investment was 17 months.

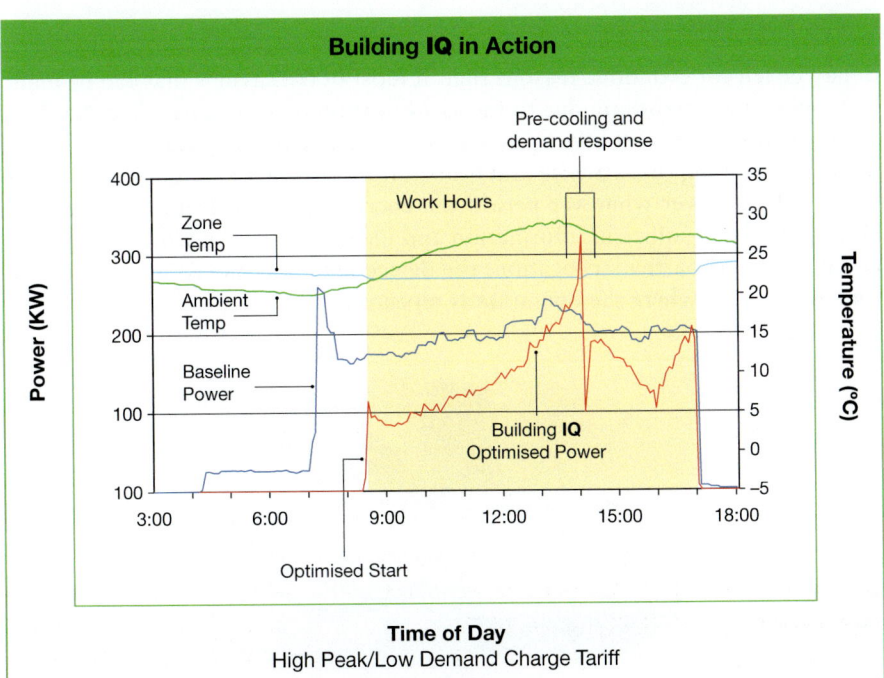

Figure 2.9 In this example it may be seen that heating was kicking in an hour earlier than needed. After optimisation, energy use was matched with building occupancy, achieving significant savings.

Source: BuildingIQ

Figure 2.10 The levels at which heating kicks in can be matched to ambient temperature.

Source: BuildingIQ

are being repeated at weekends when no one is there. Adjusting the programmer and thermostat, if relevant, to account for this would achieve significant savings at no cost to users' comfort.

Data loggers

It may be felt for economic reasons that, if most of the savings that can be made by monitoring systems are due to the identification of anomalies, then the same result could be more cost-effectively achieved with a temporary data logger. These work by clamping on to a wire and transmitting electricity use data back to the logging device over whatever period the energy manager deems to be useful. Having recorded sufficient information and identified ways in which energy can be saved at this particular location, the device is then removed and positioned somewhere else, where the operation is repeated.

Figure 2.11 A data logger clips on to any cable for on-the-spot power consumption measurement.

Source: WISPES

Data loggers may be appropriate in situations where continuous fine-grained data recording is not required, where budgets are low, and in smaller facilities. One would, however, need to keep checking back in the same locations to make sure that the previous high level of energy use has not returned for one reason or another. One may therefore cycle the logger through, for example, ten separate locations over ten weeks to fine-tune the adjustments.

Proactive systems

The above systems allow only reactive behaviour on the part of the energy manager. Because weather and energy market conditions are highly dynamic, further efficiency gains can derive from having a real-time ability to modify systems' behaviour. Energy management software is available that can predict and then automatically optimise a building's energy use, continuously adjusting its management system's settings to meet the needs of its occupants and processes each day, while delivering savings.

This approach may be suitable for large concerns that are heavy energy users, who often buy electricity or gas on a daily basis based on fluctuations and flexibility in electricity markets in order to maximise savings. Energy pricing can be volatile. Energy use can depend dynamically on many external factors, not least of which is the weather. Utilities and managers therefore often have inaccurate data as to what demand reductions in a given site or building could potentially deliver in terms of savings. The Australian Government Research Organisation (CSIRO) has tested predictive modelling and system control software in this context and found in trials that it can produce up to 40 per cent ongoing HVAC energy savings and up to 30 per cent peak load reduction. Office buildings, retail shopping malls, utility programmes and U.S. Department of Energy facilities are among the sites taking part in these trials. Among the inputs required are energy price signals, weather forecasts and predictive dynamic models of the building.

Such a system would automatically and continuously fine-tune HVAC operations to minimise energy use, connecting a building to the smart grid. It would mean that building owners, energy managers or facility managers are able to access information in real time to meet demand response energy reduction targets, while meeting the requirements of the occupants for comfort. Such innovations still require energy managers to monitor them and ensure that the solution is appropriate in every case, but by automating many tasks it leaves the manager free to engage in other activities such as encouraging employees to become more energy aware.

Person positioning systems

The reconciliation of actual energy use with predicted use can be helped by knowing the level of occupancy and use of parts of a site. Systems are now available that can allow energy managers or building owners to track individual occupants using their wi-fi-enabled mobile phones with very accurate positioning. With the permission of individuals, the information could be displayed on a real-time map showing their location and movement, or logged into a detailed occupancy database. This would make it easier to plan for HVAC setbacks, plan air-handling upgrades based on zone occupancy, prioritise lighting retrofits, make accurate predictions based on various occupancy scenarios, and much more. For example, if there is a sudden influx of people into a conference room or lecture hall, adjustments can be made to the HVAC system before the room starts to warm up.

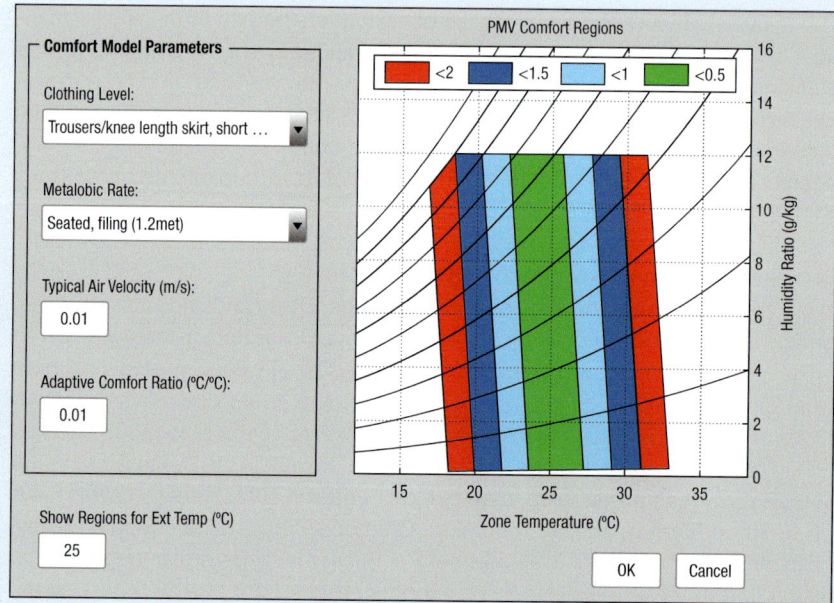

Figure 2.12 Predictive modelling can help optimise HVAC operating parameters. This building energy management system allows operators to vary the HVAC system according to building occupants' activities and clothing in relation to temperature and humidity.

Source: BuildingIQ

The Measurement and Verification (M&V) Plan

We have reached the point now where we can begin to consider a measurement and verification (M&V) Plan (as described in Chapter 1). An M&V plan is a document that defines project-specific M&V methods and techniques that will be used to determine savings resulting from specific energy efficiency projects. The plan may include either a single option that addresses all the measures installed at a single facility, or several M&V options to address multiple measures installed at the facility.

If a company is ISO 50001 certified, or aiming for certification, it will be required to implement and maintain a plan of this nature. A well-organised library of documents containing the necessary levels of detail is vital for successful compliance with this standard. In addition to providing accurate and conservative methods to calculate energy savings, a good M&V plan is clear, consistent and repeatable. It is understandable by anybody else and may be given to subcontractors or an Energy Services Company (ESCo), for example, in the form of a long-term contract. It is therefore very important to ensure that all assumptions, procedures and data are recorded properly so that they may be easily referenced and verified by others.

M&V activities include site surveys, energy measurements, metering of key variables, data analyses, calculations, quality assurance procedures and reporting. Sample specifications are provided by the U.S. Department of Energy Federal Energy Management Program, and the Carbon Trust in the UK. In general, the contents of a project-specific M&V Plan should provide an overview of the energy conservation measurement and verification activities, including the following:

- the goals and objectives, with key performance indicators, such as the energy, cost and carbon emission savings expected;
- the scope and nature of the work;
- the techniques to be used for each measure;
- the key physical characteristics of the facility, system and measure;
- the critical factors that affect energy consumption of the system or measure;
- the key baseline performance characteristics, such as lighting intensities and temperatures;
- the baseline operating conditions, such as loads and hours of operation;
- all measurements, data analysis procedures, algorithms and assumptions;
- all performance period verification activities, including the parameters to be measured, period of metering, accuracy requirements, calibration procedures, metering protocols and sampling protocols;
- the schedule for reports and procedures.

The plan will describe the source of all savings, including energy, water, O&M, and other (if applicable). There will be details of the baseline data collected, including the following:

- the parameters monitored/measured;
- the equipment monitored, i.e. location, type, model, quantity, etc.;
- the duration, frequency, interval, and seasonal or other requirements of measurements;
- personnel, dates and times of measurements.

The more comprehensive BEMS software supports M&V by including four stages of use for the gathered data:

1   Awareness: which sets a consumption baseline and can look at power quality.
2   Efficiency: which allows the operator to make incremental and proactive behavioural, control and equipment improvements, such as the load aggregation and rate analysis.
3   Optimisation: which models production using energy as an economic variable. For this stage production metrics, regulatory reports and behavioural and climate forecasts would be factored in.

4   Aggregation: which will compare consumption information against production output data and other resource planning-level information.

Models may be explored using the software to provide evidence for savings to the bottom line based on different lines of action. The energy manager would use this information to mount a case for investment in efficiency to senior management.

The remainder of this book is concerned with suggestions for how the details of the plan could be tackled, by order of topic, for a wide range of situations.

3
Airtightness and insulation

Most industrial sites will contain buildings. The energy consumed in heating and cooling a building is a function of its heat gains, the quality of the building fabric and the behaviour of the building's occupants. Heat gains arise through the following ways, which are called 'passive' heat gains.

External loads

- orientation
- window area
- glazing properties
- shading system
- insulation level (U-value or R-value: See Appendix for explanation)
- ventilation level.

Internal loads

- occupants
- lighting
- equipment.

Heat leaves or enters buildings in the following ways:

- uncontrolled draughts, via gaps in the building envelope;
- controlled ventilation;

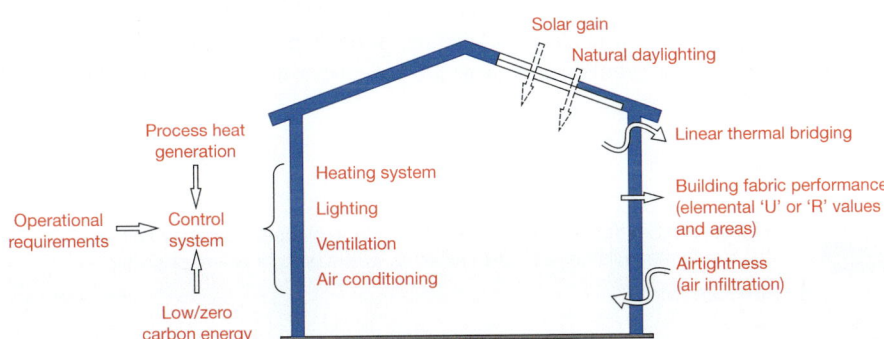

Figure 3.1
Factors affecting internal temperature.

Source: Author

- conduction through the surrounding material of the building, its roof, walls, floor, doors and windows, otherwise known as the building envelope.

Depending on the exterior temperature, this can either result in unwanted heating or cooling within the building. All of these factors need to be measured and controlled if the aim is to reduce the total amount of artificial energy input required.

Low or zero carbon buildings

Nowadays, the trend is increasingly towards low carbon or zero carbon buildings. The American Council for an Energy-Efficient Economy projects that in the medium-term future no large commercial buildings will need any heating.[1] The strategy advocated in building regulations to achieve this is to minimise the energy required to cool or heat the building through superinsulation and airtightness, at the same time as maximising the solar gain for heating, or minimising it in a hot climate to reduce air-conditioning needs. Should that strategy alone not achieve a zero energy requirement, any further energy requirements would be satisfied from heat reclamation or renewable energy, either on- or off-site.

New buildings and the refurbishment, or weatherizing (US term), of existing buildings are described below. A major weatherization of an entire building is an ideal opportunity to reduce its heating/cooling demand almost completely through superinsulation, because the additional cost of doing so will be much less over a lifetime basis. According to the US Federal Energy Management Administration (FEMA), 'the effective life of an office building is 20 to 30 years, after which major renovation and updating is normally necessarily'. Financial means of achieving this are discussed in Chapter 13.

In a superinsulated building, ventilation is controlled in a way that permits the occupiers to regulate the air's humidity, temperature and cleanliness. Ventilation can be passive, using the stack effect, if available, or active, using pumps and ducting with heat reclamation if appropriate. This means that the pumps use a heat exchanger to recover the heat from the air leaving the building and pass it into the incoming air. This is known as mechanical ventilation with heat recovery (MVHR).

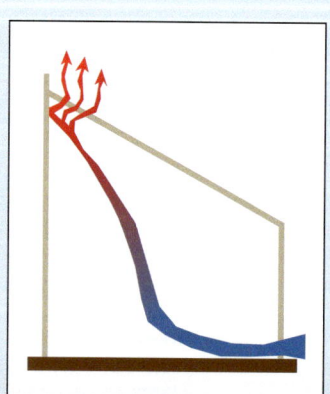

The stack effect

The 'stack effect' is used to help circulate heat in a building and moderate its internal temperature and climate. It uses the principle that warm air rises by convection to exit at the top through open windows, vents, chimneys or gaps.

Figure 3.2 Illustration of the stack effect. Warm air rises, and warm air leaving the top of the building will draw in cold air at the bottom where possible.

Source: Author

In the absence of a refurbishment opportunity, the first step to reduce heating or cooling energy demand is to remove unwanted airflow by draughtproofing.

Draughtproofing

The quickest and cheapest wins for retaining heat, or, in a hot climate, of keeping out heat, are achieved by reducing unwanted airflow in and out. A check should be made of each room, as well as the exterior. All unwanted openings must be sealed. Dampers should be fitted to any chimney flues, and intermittently running extractor fans should have well-fitting, self-closing covers which shut securely when not in use.

Badly fitting doors and windows are a major source of draughts. Small gaps around the frames can be filled with gun-applied sealants and fillers. Draught-stripping around the openings themselves is inexpensive and greatly improves comfort as well as reducing fuel bills.

(a)

(b)

(c)

(d)

Figures 3.3a, b, c and d A variety of common draught seals: (a) brush seal; (b) wiper seal; (c) compression seal; (d) service seal.

Source: (a–c) Author; (d) Chris Twinn

Windows and doors

Some doors or windows may need to be replaced. Existing single-pane windows, and windows with metal frames, should definitely be replaced. The glazing should have a low-emission (low-E) coating to allow light to pass through and reflect infrared wavelengths back in. If new windows cannot be fitted, then install secondary glazing, which fits inside the reveal and may be removed for cleaning purposes. The edges should be compression sealed. Replacement doors and windows should have insulation between the two outer surfaces to prevent thermal bridging, and preferably be constructed from timber for environmental reasons. These measures typically pay for their cost in the value of the energy saved in three to four years.

Controlling passive gain

Windows are essential to provide free lighting and heating, but this must be controlled so that occupants do not experience glare and the building does not

Left: Figure 3.4 Three levels of seal and a top-performing handle/catch system guarantee no draughts on this Passivhaus-certified door.

Below: Figure 3.5 Insulation within the frame and between the panes of a triple-glazed window or similar door removes thermal bridging and contributes to the Passivhaus level of performance.

Source: Manufacturer

Figure 3.6 Exterior
shutters are used in a
hot climate to keep
out unwanted heat
during the daytime.
They are opened in
the evening if required.

Source: Author

overheat. Solar gain is retained whenever sunlight falls on to an uncovered floor or wall that will absorb the heat within what is called its 'thermal mass'. This is the product of the specific heat capacity of the material and its total mass and conductivity. The more dense it is (e.g. stone or concrete), the more heat it will hold. Having warmed up during the day, the walls release their heat into the internal space overnight, moderating extremes of temperature.

Hotter climates

In the US region of Sacramento, California, a west-facing window of just $5m^2$ ($55ft^2$) will add as much as 16kWh (55,000Btu) to a building on a summer day. Compensating for this with an air conditioner would require almost 2kW per hour. This could be saved by the use of an exterior shade for a fraction of the cost. External shading may be provided by trees, which have an additional cooling effect. Fixed architectural elements for shading include overhangs, pagodas, vertical fins, balconies and false roofs. External shades are about 35 per cent more effective than internal ones. Overhangs are sized relative to the latitude, location and window size. Adjustable elements include awnings, shutters, blinds, rollers and curtains. Whitewashing or painting exterior surfaces a pale colour, having no skylights and roof windows, keeping shutters and curtains closed in the day and open at night, and external insulation are other tactics that can be employed.

Figure 3.7 This
pergola provides
shading to prevent
overheating. The
French windows are
Passivhaus certified.

Source: Author

Figures 3.8a and b External shading devices added to the outside of a highly glazed south-facing office block to prevent overheating in the summer months but maximise solar gain at other times of the year. Welsh government buildings in Aberystwyth.

Source: Author

Figure 3.9 The addition of glazing positioned away from the building exterior announces the sun's heat to warm up the thermal mass of the building's wall, which is then communicated inside to moderate temperature fluctuations and supplement the heating. WISE building at the Centre for Alternative Technology, Wales.

Source: Author

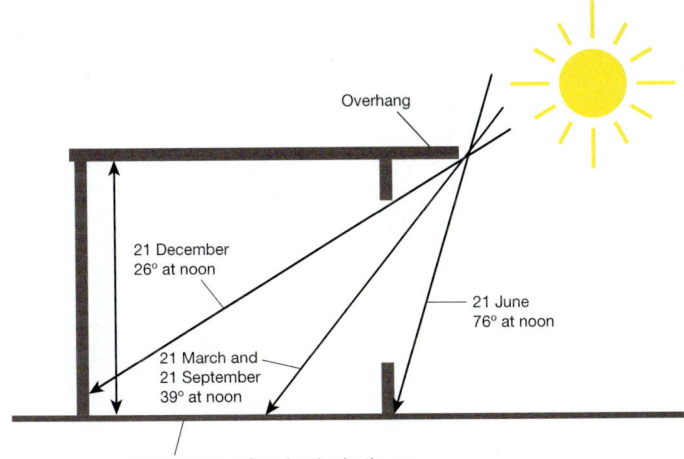

Figure 3.10 The addition of an overhang over a window can prevent glare and heat gain in the summer, but permit low-angled sun to provide light and warmth in the winter.

Source: Author

Figure 3.11 Deciduous trees may be planted outside to prevent overheating or glare in summer. In winter, without leaves, the low-angled sun may provide extra warmth.

Source: Author

Cooler climates

In cooler climates, equator-facing windows draw in heat and light. Skylights and rooflights are beneficial as long as precautions are taken to avoid overheating and to avoid heat loss at night and during cold periods. Building design can allow this direct solar heat to be conducted to other parts of the building. Automatically adjusting, motorised window-shading and insulation devices may be installed which are controlled by wireless sensors that monitor the temperature, sunlight, time of day and room occupancy. These devices can reduce the energy requirement for cooling by over half.

Superinsulation and airtightness

Introducing airtightness and superinsulation, whether in a cold or a hot part of the world, will maximise the potential for control over the energy cost. This is achieved by installing a continuous airtight layer around the building envelope, to prevent unwanted draughts through gaps in the building fabric, and thick layers of insulation.

Figure 3.12 Poor building practice like this leads to air gaps and thermal bridges that would later be covered up and be invisible to a visual inspection. External insulation is the simplest way to rectify such problems.

Source: Stamford Brook report on airtightness, courtesy Malcolm Bell, Leeds Metropolitan University, reproduced in EST (2008)

Figure 3.13 Detailing showing the position of an airtightness barrier (blue dotted line) beneath the insulation under a suspended floor, lapping up the side to meet plaster/plasterboard. There is further insulation on the inside of the plaster. Insulation is also on the outside of the building. Together this makes the building thermally insulated and airtight. Diagrams like this are free from the enhanced construction details (ECDs) section of the Energy Saving Trust website.

Source: EST

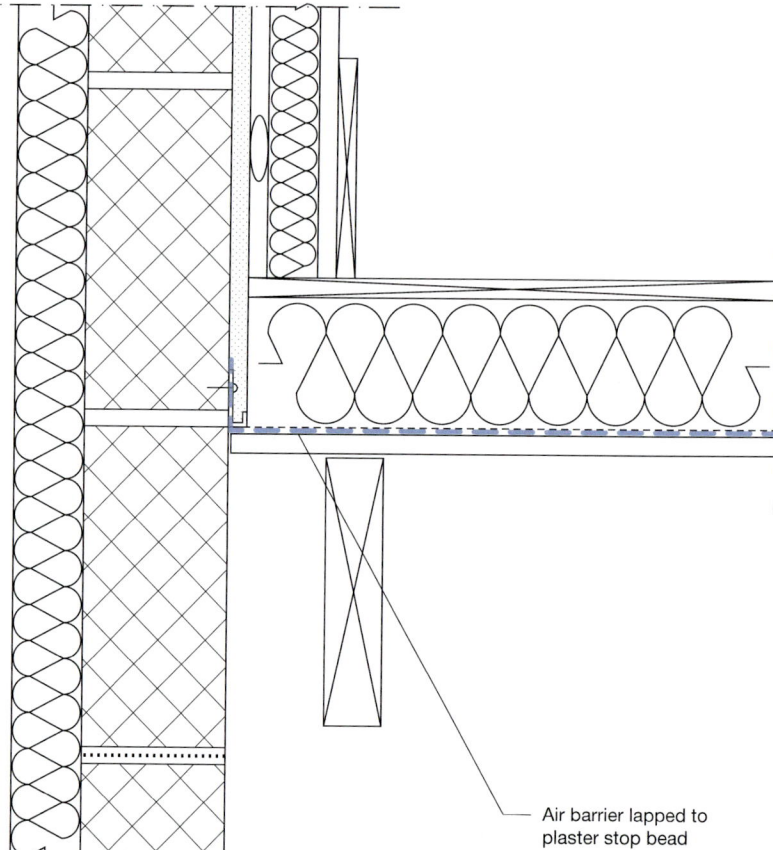

Air barrier lapped to plaster stop bead

What can the airtightness barrier consist of?

Vapour permeable (and hygroscopic) materials

Vapour permeable membranes, intelligent membranes, lime, concrete, timber, hemcrete, bricks, stone, plaster, mineral (rock and slag) wool (Rockwool).

Non-permeable materials

PVA and vinyl paint, when used as a coating on plaster, metal, glass, PVC, plastic, plastic foams, XPS, EPS, phenolic foam insulants.

If problems with damp and condensation are present or anticipated, moisture must be able to pass through this layer; this is not possible with non-permeable materials. Instead, the materials used should be hygroscopic and vapour permeable. If most of the exterior building skin is impermeable, as in the case of buildings clad in sheet metal or insulated with polystyrene or foam-based insulants, condensation problems may be resolved with the use of a dehumidifier in the HVAC system, but this adds additional energy load.

Professional long-lasting tape should be used to seal joins in airtight membranes. The devil is always in the detail and the most problematic areas are where the wall or roof is penetrated by windows, skylights, joists, doors and service entries. The airtight barrier must be able to withstand pressures created by wind, stack effect or ventilation systems. For large structures, every floor or occupancy unit should be treated as its own airtight area.

How airtight should it be? Building regulations specify the number of air changes per hour permitted for safety. The introduction of fresh air is managed with controlled ventilation. The level of airtightness is tested by professional pressure testing devices, called blower doors. The common metric is the number of air changes per hour at a specified building pressure, typically 50Pa (otherwise known as ACH50, where ACH stands for air changes per hour; a Pascal is equivalent to one newton per square metre (N/m^2)). The standard of airtightness for Passivhaus certification (see below for more information) is that a new building must not leak more air than 0.6 times the building volume per hour at 50Pa. This standard is around four times better than most building regulations. A refurbished building could achieve the EnerPHit standard, which is designed to make such buildings close to Passivhaus but reduce costs. With this, the criteria have been relaxed a little so that the airtightness level is one building volume air change per hour.

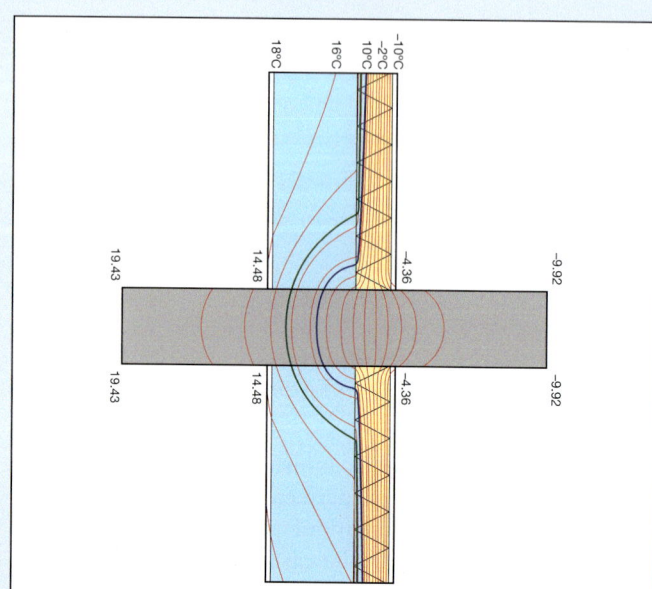

Thermal bridging

Thermal bridging occurs when a relatively conductive material passes through the building envelope from the interior to the exterior. This may be a single-paned window, or a concrete lintel in a windowsill, any fixings, window or door frames, joists, services and wall ties in cavities. Sometimes the floor slab is extended through the building envelope. As the standard of airtightness and insulation increases in a building, thermal bridging becomes increasingly significant as a factor in heat transfer. To 'break' a thermal bridge, insulation is added on the inside or outside, or even in between, as seen in the door and window designs earlier in this chapter.

Figure 3.14 A concrete beam passing through from the inside (left, at 18°C/65°F) to the outside (right, at –10°C/14°F), first through insulation, and then through timber. The concrete beam acts as a thermal bridge passing both heat from the inside outside, and cold from the outside inside. The temperature gradient is modelled in software.

Source: Wikimedia Commons. Author: Bauigel

Insulation

Commercial and industrial premises come in a huge variety of sizes and shapes. They also come constructed from many different materials, from traditional brick and stone, to modern steel warehouses. Retrofitting insulation on to these premises requires different strategies for each building type.

The highest possible standard of insulation should be aimed for, but there is a law of diminishing returns. The final standard is determined by budget, the space available and choice of material. It is vital to avoid any gaps in insulation. A detailed discussion of building insulation and installation strategies is in the companion volume *Sustainable Home Refurbishment*.

Choice of insulation material

There are three recommended, economic types of insulating material for a low carbon building: mineral wool (e.g. Rockwool), wood fibre and recycled cellulose (e.g. Warmcel). The first two, like foam insulants, come as boards and batts. The latter is loose fill and only proven to be suitable for horizontal spaces when

U- and R-values

The U-value of a material is a measure of its ability to resist heat loss. The lower the U-value a material has, the better insulation it provides. In the USA, R-values are used instead of U-values. They are the inverse of U-values. Therefore, the higher the R-value a material has, the better insulation it provides. To convert R-values to U-values, divide into 1. The R-value of the building is calculated by averaging the sum of the R-values of all elements multiplied by the area of the building's external surface (see the Appendix for more information on these, and k-values).

contained and uncompressed, in which case it is easy to install, cheap and highly effective.

There is a trade-off between the depth of insulation required to achieve a given level of insulation value, and their impact on the environment, as measured in embodied carbon. Natural materials such as wood fibre and recycled cellulose (newsprint) act as carbon sinks, locking the atmospheric carbon which the plants absorb during their lifetime into the fabric of the building. EPS/XPS/Phenolic foam materials not only do not breathe, but emit carbon dioxide into the atmosphere during their manufacture. Table 3.1 shows how this trade-off works with

Figures 3.15a and b Tongue-and-groove wax-impregnated wood fibreboard cladding over wood fibre batts. The cladding is rendered with lime plaster to make the walls breathable, which reduces the risk of condensation inside.

Source: Author

Table 3.1 Four common insulants compared. The higher the figure for embedded carbon, the worse the impact on climate change. A negative figure means a positive impact.

Insulant	k-value (W/mK)	Level of insulation to achieve a U-value of 0.15W/m²K	Climate change impact (embodied carbon in kgCO₂e)
EPS/XPS/Phenolic foam	0.020–0.025	130mm	160
Mineral wool	0.033–0.40	225mm	20
Recycled cellulose	0.038–0.040	250mm	−1.9
Wood fibreboard	**0.080**	**500mm**	**−35**

four common insulants. Interlocking panels of wood fibre are available specifically for exterior cladding, which eliminates thermal bridging. When lime rendered, they are weather resistant and breathable. At the end of their life they are compostable.

External or internal?

A decision needs to be taken first on whether external or internal insulation should be applied. In general, internal wall insulation is cheaper than external insulation. On the other hand, there will be a loss of internal space, it is disruptive, and it is necessary to move electrical sockets and light switches, skirting boards and so on. The thermal heat storage value of the wall is lost, so it will heat up faster, but it will also cool down quicker.

External insulation improves weather protection and noise insulation as well as retaining the thermal mass of the walls inside. Any value of insulation may be achieved and there is little inconvenience to occupants during installation. Any gaps and cracks in the wall or poor rendering are covered up. It is easier to ensure airtightness. Detailing around windows and doors is more easily managed, and it becomes possible to insulate a whole building at once, with lower overheads on scaffolding, etc. Choosing external insulation for a thermally massive structure also implies that the best choice of heating is a constant and low-level one, such as underfloor heating.

External systems

External insulation involves applying an insulating layer and a decorative weatherproof finish to the outside wall. The aim is to reach U-values below 0.3W/m²K or half of this for the Passivhaus standard (see below). Modern façade cladding systems for large and high-rise buildings are often made off-site and installed quickly on brackets fixed to the wall. This maximises the possibility of airtightness. A steel frame holds the insulation, and the coloured render is added either before or after fixing. They can be ordered to specified levels of noise and thermal insulation.

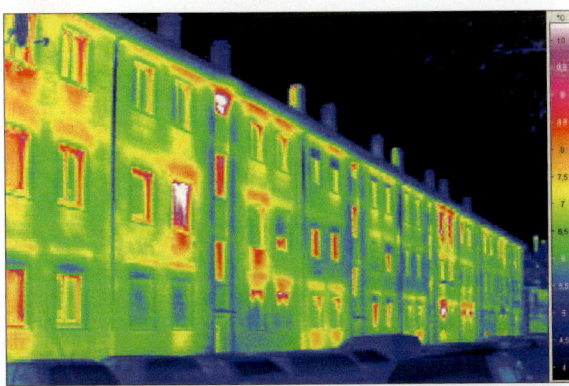

Figures 3.16a, b and c External insulation on terrace apartments in Frankfurt, Germany, with thermographic photographs taken before and after modifications. Blue indicates that there is no heat escaping. Thermographic photography is an excellent a way of revealing heat loss. The design was calculated using PHPP software, which is necessary when designing Passivhaus standard buildings. It achieved an annual heat energy demand of 17kWh per metre-squared area. The strategy was 260mm exterior insulation, triple glazing, central ventilation heat recovery, reduction of thermal bridges and 7.5m^2 (80ft^2) of solar collectors. A pressure test recorded an air change rate of 0.461/hr.

Source: International Energy Agency

Choice of render

In terms of renders and their environmental impact, hydraulic lime renders are preferred, since they provide a lasting, breathable surface. They consist of around 12mm coat of hydraulic lime containing glass fibre mesh reinforcement topped with a 3mm decorative layer. Acrylic renders, consisting of two coats of 4 to 6mm reinforced with mineral fibre or glass mesh, are the next best performing, and are likewise highly elastic, weather resistant and breathable.

Internal insulation

There are two techniques for internal insulation: insulated plasterboard applied directly on to the internal walls, or studwork. The first of these techniques will achieve a higher insulation the greater the depth. Airtightness is preserved by applying a continuous band of adhesive to it around the edges of the wall and all openings such as sockets and plumbing. The best boards include a vapour control layer to stop moist internal air condensing on the cold break behind the insulation.

Studwork is employed on a wall that has previously suffered from damp or where the surface is not in a strict plane. The studwork acts as a thermal bridge, which must be broken by insulation on the front. Alternatively, insulating studs, which are made of extruded polystyrene laminated to oriented strand board, are available. An intelligent vapour control layer is fitted over this by continuously lapping and bonding the membranes together, sealing them to the floor, internal walls and windows. This acts as the airtightness barrier. At all costs, steps must be taken to avoid puncturing this layer.

Floors

In the case of a solid ground floor, insulation should be applied over the top. Beneath this the vapour control layer is applied, which laps up the sides all around to join the layer coming down the wall. The floor surface is laid on top of the insulation.

Roofs

The type of roofing insulation depends on whether the roof is flat or pitched. A U-value of at least $0.23W/m^2K$ should be aimed for. The insulation in a flat roof

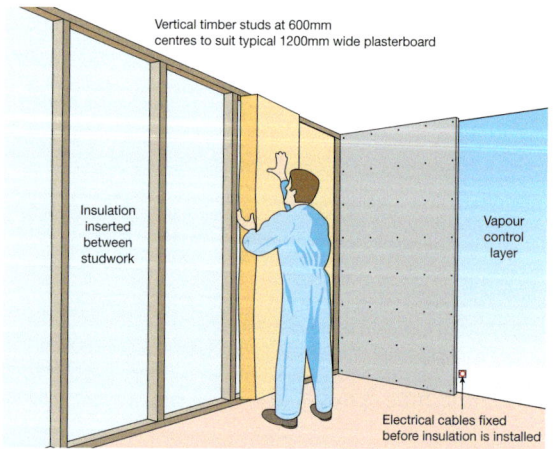

Figure 3.17 Installing internal insulation, dry lining style.

Source: EST

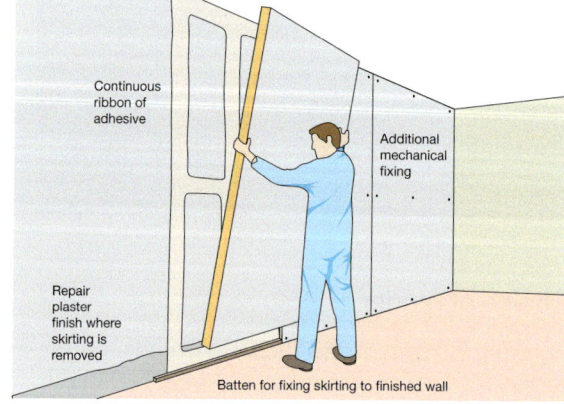

Figure 3.18 Installing internal insulation by sticking it directly to the wall.

Source: EST

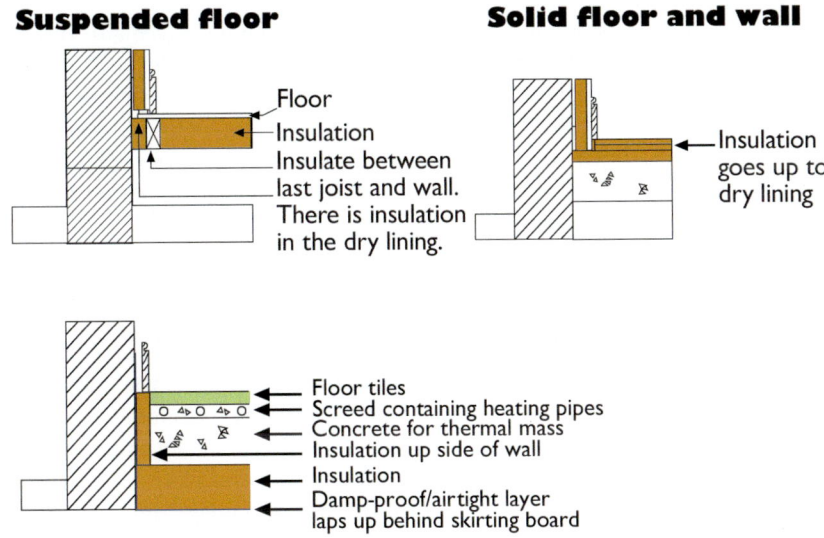

Suspended floor

Floor
Insulation
Insulate between
last joist and wall.
There is insulation
in the dry lining.

Solid floor and wall

Insulation
goes up to
dry lining

Floor tiles
Screed containing heating pipes
Concrete for thermal mass
Insulation up side of wall
Insulation
Damp-proof/airtight layer
laps up behind skirting board

Figures 3.19a, b and c Strategies for underfloor insulation. All of these strategies will minimise thermal bridging. The strategy shown in 3.19c includes underfloor heating, which is recommended when a floor is being replaced, as it is the most energy-efficient form of heating.

Source: Author

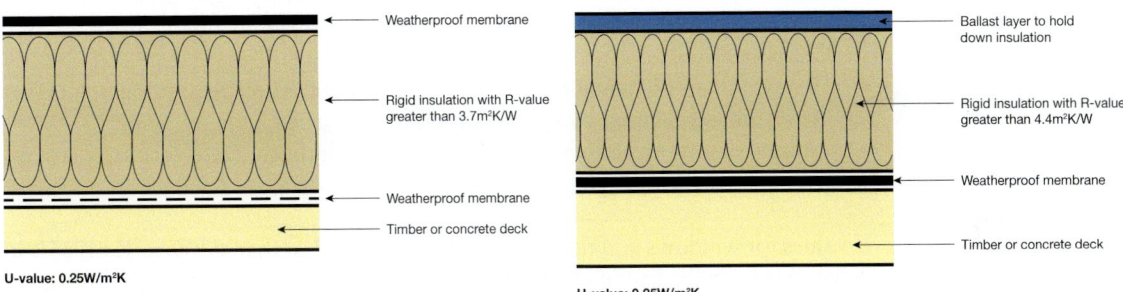

Weatherproof membrane

Rigid insulation with R-value greater than 3.7m²K/W

Weatherproof membrane

Timber or concrete deck

U-value: 0.25W/m²K

Ballast layer to hold down insulation

Rigid insulation with R-value greater than 4.4m²K/W

Weatherproof membrane

Timber or concrete deck

U-value: 0.25W/m²K

Figures 3.20a and b Warm roof (a) and cold roof (b) flat roof insulation.

Source: EST

should ideally be located between the roof deck and the weatherproof membrane in a warm roof deck construction. Careful detailing at the edge and parapet areas of flat roofs is vital for reliability and durability.

For a pitched roof the aim is 0.16W/m²K. Here, it can only be installed externally if the roof is being replaced. Otherwise, internal roof insulation is applied in a similar way to dry lining on vertical walls. There must be no gaps between the slabs. At the base of the sloping ceilings, purpose-made eave vents should be installed that provide the equivalent of a 25mm or one-inch continuous gap, as well as ventilation at the roof's ridge in order to cross-ventilate the roof space and prevent condensation.

Passivhaus and EnerPHit standards

The Passivhaus standard (see www.passiv.de) is the most reliable method for achieving low and zero carbon buildings for new and existing structures, and has been proven with thousands of examples. It stipulates a way of analysing all of the heat gains and losses within a building, and modelling improvements through the use of software to achieve the optimum result cost-effectively.

The standard specifies that the space heat requirement must not exceed $15 kWh/m^2/yr$. Total primary energy use (of all appliances, lighting, ventilation, pumps, hot water) must also not exceed $120 kWh/m^2/yr$ ($38039 Btu/ft^2/yr$). Building fabric U-values must be less than $0.15 W/m^2 K$. This implies an airtightness level of 0.6 times the building volume per hour at 50Pa (N/m^2), as mentioned above. A lower standard for a refurbished building is the EnerPHit standard, where the space heating energy requirement is $25 kWh/m^2/yr$, with an airtightness level of one building volume air change an hour.

The strategies available to achieve these standards address thermal bridging, airtightness and all energy use. Designs are modelled using the Passivhaus Planning Package (PHPP) software, which may be purchased from the Passivhaus website. The requirements imply the following features to achieve them:

- Passive preheating of fresh air: brought in through underground ducts that exchange heat with the ground to reach above 5°C (41°F), even on cold winter days (see Chapter 6).
- MVHR (Mechanical Ventilation with Heat Recovery): transfers over 80 per cent of the heat in the ventilated exhaust air to the incoming fresh air (see Chapter 5).
- Hot water supply using renewable energy: solar collectors, biomass, CHP or heat pumps powered by renewable electricity (see Chapters 5 and 6).
- Energy-saving appliances: ultra-low-energy lighting, refrigerators, stoves, freezers, washers, dryers and so on.

U-values for windows and doors generally need to be less than $0.8 W/m^2 K$ ($4755 Btu/ft^2/yr$) (for both the frame and glazing) with solar heat-gain coefficients around 50 per cent. All passively designed buildings deserve to win their plaudits based on proven performance in use rather than through a post-completion calculation.

Enhanced construction details

It is worth mentioning a free tool that is very useful in some circumstances. Enhanced construction details (ECDs) that focus on heat losses that occur at the junctions between building elements (walls, ceilings, floors) and around openings are freely available from Britain's Energy Saving Trust. They are designed to help the construction industry to maintain high performance standards and were developed in association with an industry working group. Using the complete set of three ECDs and ensuring that all remaining details achieve regular standards will obtain a thermal bridging λ of $0.04 W/m^2 K$. See: http://bit.ly/Qsbs2r.

Note

1 John A. 'Skip' Laitner, Steven Nadel, R. Neal Elliott, Harvey Sachs and A. Siddiq Khan. 2012. *The Long-Term Energy Efficiency Potential* (http://www.aceee.org/ research-report/e121). Washington, DC: American Council for an Energy-Efficient Economy.

4

Lighting, daylighting and controls

As a proportion of overall energy use lighting can figure surprisingly large, being responsible for up to 30 to 40 per cent of an electricity bill in sites that have not already embraced energy-efficient lighting. Worldwide, lighting accounts for over 19 per cent of total electricity consumption.

There are usually plenty of opportunities for reducing the cost of lighting by up to 90 per cent if best practice is followed with the use of daylighting being maximised and conventional lighting converted to LEDs. This should simultaneously improve the quality of illumination.

Building regulations in different countries specify certain levels of lighting and efficiency for different purposes. In California, for example, the Energy Independence and Security Act of 2007 requires that commercial building owners must ensure that lighting uses less than 1.1 watts per square foot. This means that bulbs must have 25 per cent more efficacy than the standard 100-watt

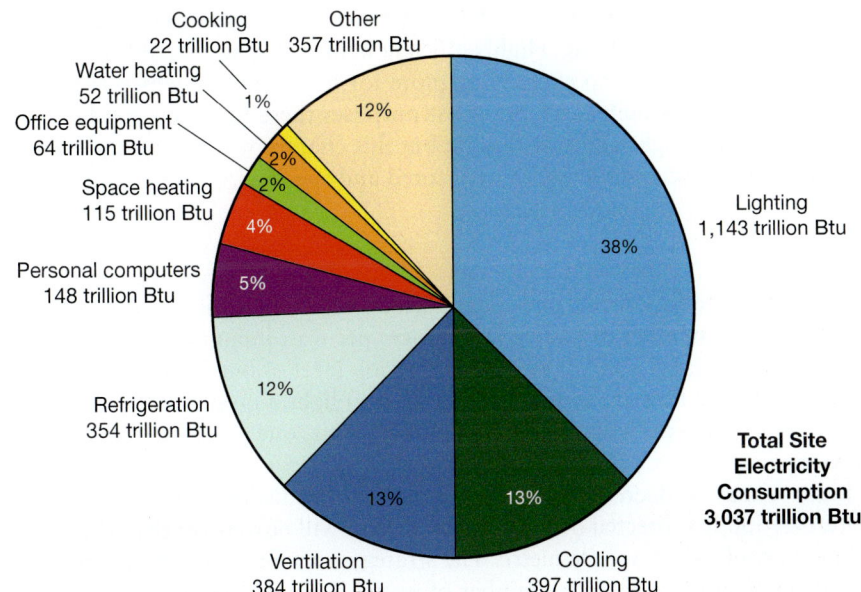

Figure 4.1 In 2009, lighting used more power than any other application in American industrial and commercial/office buildings.

Source: US Energy Information Administration

Figure 4.2 Efficiency gains for selected commercial equipment in three cases, 2040 (percent change from 2011 installed stock efficiency). Lighting clearly holds the most potential.

Source: EIA Annual Energy Outlook 2013

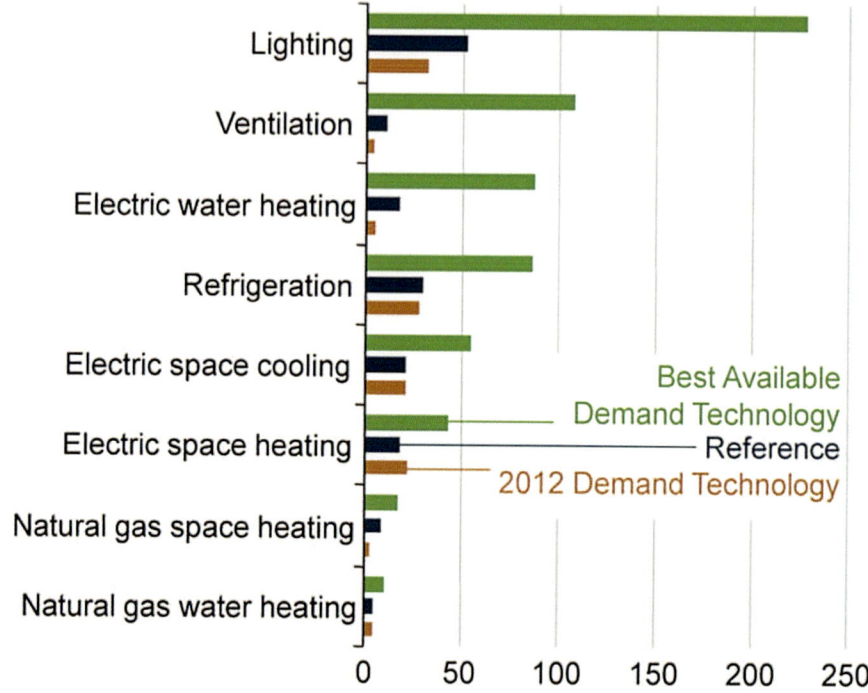

incandescent bulb nowadays. They will be just as bright (1,600 lumens) but will consume no more than 72 watts. However, the best available technology is way more efficient than this. In an industrial context, lighting may be used for a great variety of purposes, from precision lighting for delicate operations to exterior floodlighting and spotlighting. Highly efficient lights are now available for all of these purposes. There is no excuse any more for using lighting which gives off heat and is therefore inefficient. For most purposes these functions can be provided by LEDs, or light-emitting diodes; but this chapter will compare all types of lighting, and discuss how light is measured and specified.

Light

Light is measured in lumens, and illumination in lux. The efficacy of a lamp is measured in the number of lumens it produces per watt input (Table 4.1).

A lux (symbol: lx) is equal to an illumination level of one lumen per square metre. Lux reveal how many lumens you need to light a given area. One lux is equal to one lumen per square metre. (In non-SI units, one footcandle is equal to approximately 10 lux.) (Table 4.2.)

Light intensity decreases by the square of the distance from the bulb. Therefore, 500 lux directed over 10 square metres will be dimmer than the same amount spread over 1 square metre. The strategy is to maximise the number of lumens obtainable for the least number of watts. If a space, which requires 500

Table 4.1 The performance of typical 12V lamps

Lamp type	Rated watts (W)	Light output lumens (lm)	Efficacy (lm/W)	Lifetime (hours)
Incandescent globe	15	135	9	1,000
Incandescent globe	25	225	9	1,000
Halogen globe	20	350	18	2,000
Batten-type fluorescent (with ballast)	6	240	40	5,000
Batten-type fluorescent (with ballast)	8	340	42	5,000
Batten-type fluorescent (with ballast)	13	715	55	5,000
PL-type fluorescent (with ballast)	7	315	45	10,000
LED lamp (see note)	3	180	30–100	>50,000

Source: Manufacturers' data

Note: The performance of LEDs varies considerably according to the manufacturer. Choosing the right LED products is very important.

Table 4.2 The number of lux needed for different applications

Lux level	Area or activity
20–30	Car parks, roadways
<100	Corridors, stores and warehouses, changing rooms and rest areas, bedrooms, bars
150	Stairs, escalators, loading bays
200	Washrooms, foyers, lounges, archives, dining rooms, assembly halls and plant rooms
300	Background lighting (e.g. IT office, packing, assembly (basic), filing, retail background, classrooms, assembly halls, foyers, gymnasium and swimming pools, general industry, working areas in warehouses)
500	General lighting (e.g. offices, laboratories, retail stores and supermarkets, counter areas, meeting rooms, general manufacturing, kitchens and lecture halls)
750	Detailed lighting (e.g. manufacturing and assembly (detail), paint spraying and inspection)
1,000	Precision lighting (e.g. precision manufacturing, quality control, examination rooms)
1,500	Fine precision lighting (e.g. jewellery, watchmaking, electronics and fine working)

Source: Carbon Trust and lighting manufacturer Veelite.

Figures 4.3a and b
Illumination decreases by
the inverse square law with
distance from the light source.
As an example, if a bulb gives
off 400 lux at 1m, at a
distance of 4m the irradiation
will be one-sixteenth of this,
or 25 lux (reading from the
graph). If 300 lux is required
at 4m distance, then 12 lamps
each giving off 400 lux would
be required (300/25). This
illustrates the importance
of positioning in lighting. r =
radius, so 2r is 2 × the radius
shown (distance from light
source).

Source: Author & Wiki Commons,
author: Borb

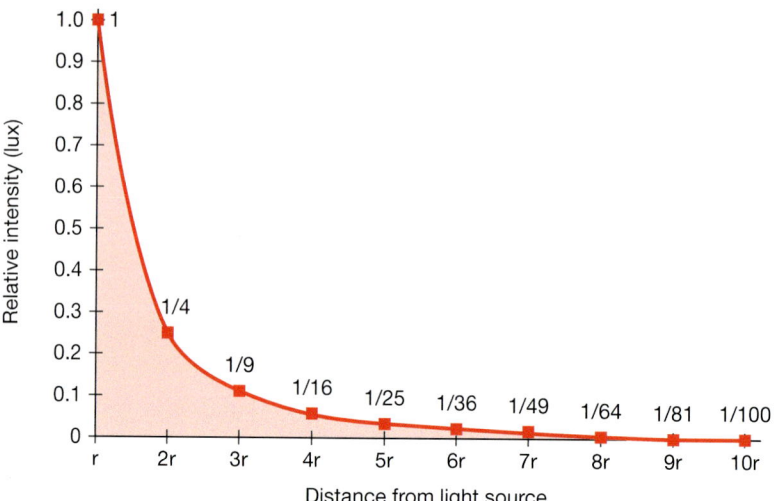

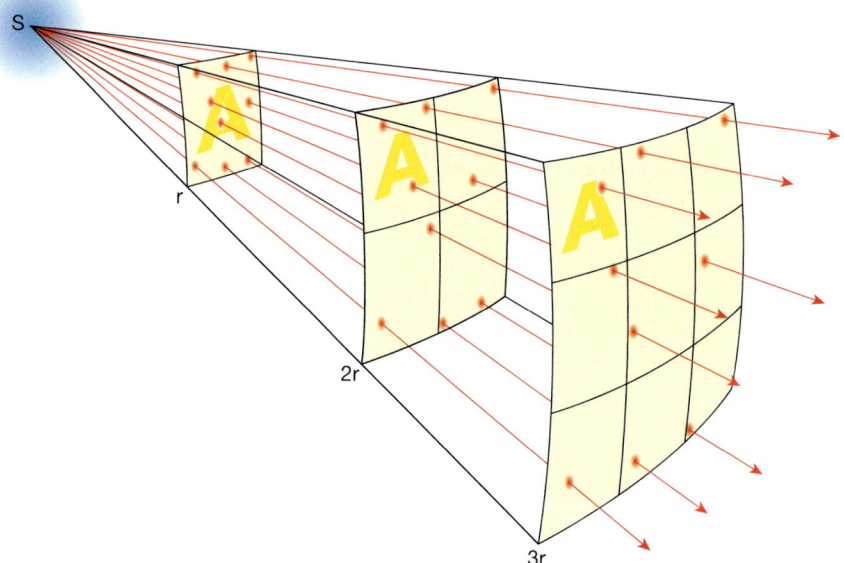

lux, has an area of 100 square metres, and the lamps are 2 metres above the bench level, then this will need 500 × 100 × 4 = 200,000 lumens. This space could therefore be lit (at night) by (using the figures in Table 4.2): 200,000/70 = 2,156W of CFLs, 200,000/90 = 2,224W of LEDs, or 2,000W of high-frequency fluorescent tubes.

Using natural light

Now let us consider daylighting. Visible transmittance (VT) is a fraction of the visible spectrum of sunlight (380 to 720 nanometers), weighted by the sensitivity of the human eye, that is transmitted through the glazing of a window, door or skylight. A product with a higher VT transmits more visible light. VT is expressed as a number between 0 and 1. This is not to be confused with the solar heat gain coefficient (SHGC), which is the proportion of total solar energy that enters via windows. Light-to-solar gain (LSG) is the ratio between the SHGC and VT. It provides a gauge of the relative efficiency of different glass or glazing types in transmitting daylight while blocking heat gains. The higher the number, the more light is transmitted without adding excessive amounts of heat. This figure is used to compare the different windows' potential performance in a given space.

Building occupants feel more comfortable and have a higher degree of well-being in sensitively lit interiors, which means maximising the use of natural light. Light directly affects mood and alertness as well as productivity, as borne witness by Seasonal Affective Disorder (SAD).

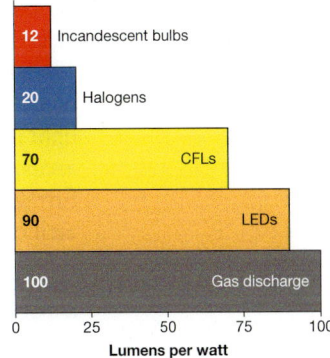

Figure 4.4 The average efficacy of different types of lighting.

Source: Author

Passive daylighting

Even existing buildings can be adapted to make better use of natural lighting. In Chapter 3 we looked at some ways in which windows may be used to admit natural light without causing too much glare to occupants. Having bright reflective surfaces and light surface colours will aid in the distribution of daylight in rooms, but this is by no means an exhaustive list.

Passive daylighting is a system of collecting sunlight using static, non-moving and non-tracking systems such as windows, light shelves, skylights, rooflights or

Figure 4.5 Daylighting should be maximised with the help of windows, but its management can be separated from that of heat gain, which may not always be required. Shading will help control unwanted glare.

Source: IEA

atrium spaces, tubular daylight devices, solar shading devices, daylight-responsive electric lighting controls, and daylight-optimised interior design (furniture, etc.). The intention is to direct low daylight high into a space (to reduce the likelihood of excessive brightness). Where possible, ceilings can be sloped to direct more light inward. It is vital to prevent direct daylight from reaching critical visual task areas, and so it needs to be filtered. Artificial light should be brought in gradually further within spaces, so that there is not a sudden contrast between natural and artificially lit areas.

Directing daylight into a building

Skylights are one way of achieving this. They can be either passive or active. Passive skylights simply let daylight enter through glazing in the roof. Active skylights contain a mirror system that tracks the sun across the sky to reflect and direct it where needed. Optionally, systems are available which reduce daylight during summer months and assist with cooling and ventilation.

A light shelf is an externally positioned architectural element that allows daylight to penetrate deep into a building. They are effective on south façades but often ineffective on east or west elevations of buildings. A horizontal, light-reflecting overhang is placed above eye level and has a high-reflectance upper surface to reflect daylight on to the ceiling. They are commonly made of an extruded aluminium chassis system and aluminium composite panel surfaces. Light shelves and louvred systems make it possible for daylight to penetrate the space up to four times the distance between the floor and the top of the window. They are generally used in continental climates and not in tropical or desert climates due to the intense solar heat gain.

Exterior louvre systems, which are vertical or horizontal fins whose opening and closing mechanisms, rather like Venetian blinds, can be controlled automatically, prevent too much sunlight from reaching a window, or can have reflective surfaces to direct light inside while avoiding glare.

Architectural light wells direct daylight further into (mostly new) buildings. Light can be brought into spaces in other ways too: by using atria and light guidance systems. Light tubes, also called sun pipes, solar pipes or daylight pipes, have reflective inner surfaces to take light from a roof level deep into lower levels. Compared to conventional skylights and other windows, they offer better heat insulation properties and more flexibility for use in inner rooms, but less visual contact with the external environment.

Figure 4.6 Plan view of a windowed wall showing adjustable louvre positioning on the outside to prevent glare or reflect deeper light into a building.

Source: Author

Operable shading and insulation devices

A design with too much equator-facing glass can result in excessive heating, or uncomfortably bright living spaces at certain times of the year, and excessive heat transfer on winter nights and summer days. Control

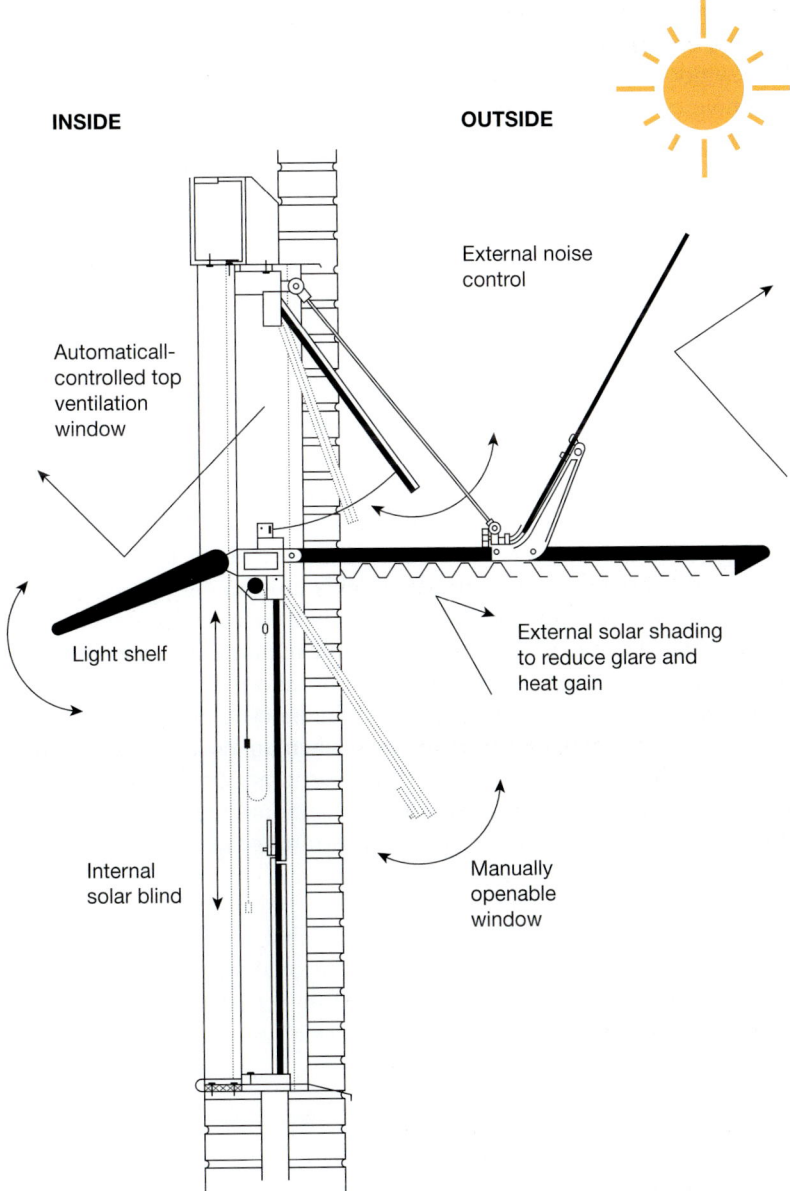

INSIDE

OUTSIDE

External noise
control

Automaticall-
controlled top
ventilation
window

Light shelf

External solar shading
to reduce glare and
heat gain

Internal
solar blind

Manually
openable
window

Figure 4.7 This
cross section
illustrates various
possible types
of shading and
light-directing fixtures
for and around
equator-facing
windows to maximise
daylighting for interiors
while controlling glare
around the year and
through the day.

Source: Author, adapted
from Mr7uj

mechanisms (such as manual or motorized interior insulated drapes, shutters,
exterior roll-down shade screens, or retractable awnings) can compensate for
differences caused by thermal lag or cloud cover, and help control daily/hourly
solar gain requirement variations. Automated systems that monitor temperature,
sunlight, time of day and room occupancy can precisely control motorised
window-shading and insulation devices. A successful lighting design that incorporates any of these features must be integrated with the artificial lighting system
using advanced lighting controls, and with the building energy management
system (BEMS) if there is one. Three types of controls are available:

1   Those which turn lights off when there is ample daylight.
2   Control of individual lamps in zoned circuits that are progressively further from windows, to provide intermediate levels of light.
3   Dimming controls that continuously modulate the power to lamps to complement the amount of daylight.

Windows

Windows consist of two or three layers of glazing. Glazing can be coated or laminated to improve its strength or optical and insulation properties.

Coatings

When specifying glazing, or windows, it is possible to order glazing with different coatings that not only admit and retain infrared heat, but also admit as much light as is required. Manufacturers supply a huge range of coatings: some permit only 6 per cent of light to enter the building, or 8 per cent of heat. Panes specifically designed to reduce energy use have extra-clear outer layers, letting up to 80 per cent of light and 71 per cent of the sun's heat in.

'Smart' windows are also becoming available, based on electrochromism, a technology that can control light and heat while maintaining view and reducing glare. The glass can be clear, opaque, tinted or coloured, has the capability to modulate heat and light transmission, and may be used in a variety of applications. The windows' optical properties vary when an electric field or current is applied across the device. This causes the absorptance or the reflectance of the active layers to change, thereby modulating the amount of light (electromagnetic radiation) that passes through the coated substrates or glass.

Energy labels

Modern windows are sometimes rated by national bodies and come with a declaratory label. In the UK this is the British Fenestration Rating Council (BFRC). This label displays the following information:

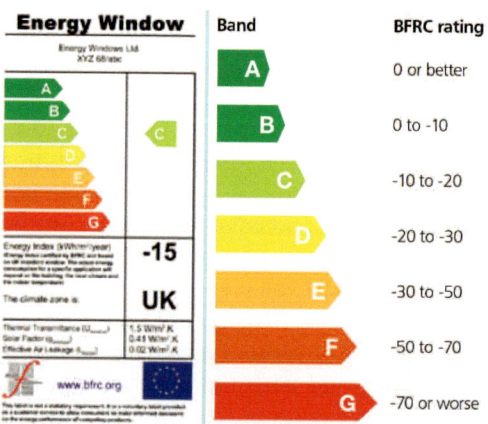

Figure 4.8 A British Fenestration Rating Council window energy rating label.

Source: British Federation Rating Council

1 the rating level: A, B, C, etc.;
2 the energy rating, e.g. $-3\,kWh/m^2K/yr$ (= a loss of 3 kilowatt-hours per square metre per year);
3 the U-value, e.g. $1.4\,W/m^2K$;
4 the effective heat loss due to air penetration as L, e.g. $0.01\,W/m^2K$;
5 the solar heat gain G-value, e.g. 0.43 (this is explained more fully in the companion volume, *Energy Management in Buildings*).

The ratings are ranked according to the amount of thermal transmittance (heat transfer) that is permitted by the window, as measured by the amount of energy (in kilowatt-hours) lost per year through one square metre of the window, divided by the difference in temperature between the inside and outside ($kWh/m^2K/yr$) (Table 4.3).

In the USA, the equivalent label is produced by the National Fenestration Rating Council (NFRC). Its Component Modeling Approach (CMA) Product Certification Program enables whole product energy performance ratings for commercial (non-residential) projects. It uses the following components:

Table 4.3 Meaning of the BFRC window energy rating label

Rating	Energy lost per year (kWh/m^2K)
A	0 (no energy lost) or better
B	0 to −10
C	−10 to −20
D	−20 to −30
E	−30 to −50
F	−50 to −70
G	−70 or worse

Figure 4.9 A sample window rating label from the National Fenestration Rating Council (NFRC). It gives the solar heat gain coefficient (SHGC) for the glazing, the U-Factor (the same as the R-Value, or inverse of the U-value), air leakage rate and its visible transmittance (VT).

Source: NFRC

Figure 4.10 Another sample American window efficiency rating label.

Source: ENERGY STAR

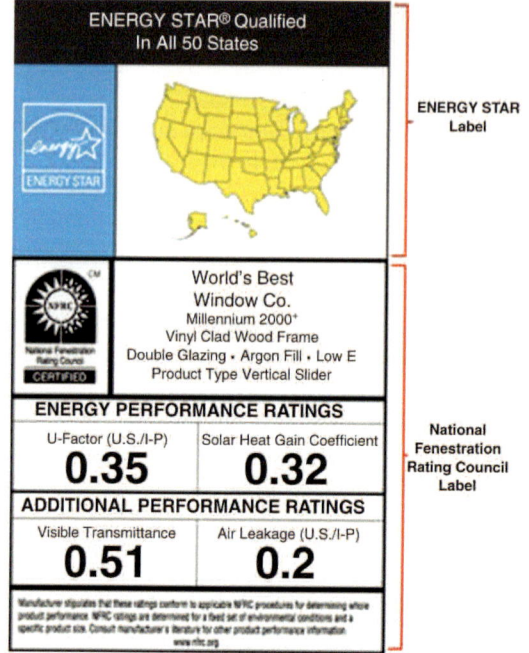

- **Glazing:** glazing optical spectral and thermal data from the International Glazing Database (IGDB).
- **Frame:** thermal performance data of frame cross sections.
- **Spacer:** Keff (effective conductivity) of spacer component geometry and materials.

Most US states' building energy codes reference NFRC 100 and 200 for fenestration U-factor and SHGC because they are required by ASHRAE 90.1, Section 5.8.2. California's Title 24 Building Energy Efficiency Standard now requires CMA label certificates for site-built fenestration in large projects.

Obtaining the label certificates is voluntary. The CMA programme does not label each product. Instead, NFRC lists the values for rated products on a single document for the entire project, called the label certificate. CMA may be used to rate commercial windows, doors, skylights, curtain walls and store fronts. To obtain a label certificate, the design team uses pre-approved, NFRC-rated components (frames, glazing and spacers) to configure a product in the CMA Software Tool (CMAST), which then generates energy performance ratings for the whole product. An NFRC-Approved Calculation Entity (ACE) then certifies the ratings and issues the label certificate.

Glow-in-the-dark concrete

Figure 4.11 German-based manufacturer Kann has developed luminous concrete pavers called NightTec Leuchtsteine.

Source: KANN GmbH

Free outdoor lighting for pedestrian access, etc. can be provided with luminous paving. Charged with sunlight by day, paving stones are available that deliver safe pathway illumination using phosphorescent crystals embedded in the surface. They capture diurnal energy and emit light every evening for about ten hours. The slabs appear white in sunlight, but at night their glow is either green or blue, depending on the selected coating. Although they only emit low-level illumination, this can be more than adequate for ground visibility in dark environments. The same technology may be applied as an acrylic render to the surface of buildings and other structures, inside and out, to save on ambient lighting costs and emissions.

Lamps

From maximising daylighting, we move to the choice of lamp, which is to say light bulb, together with its holder and shade or cover. Lamps are evaluated according to their efficiency, lifetime and colour. There are four types of lamp:

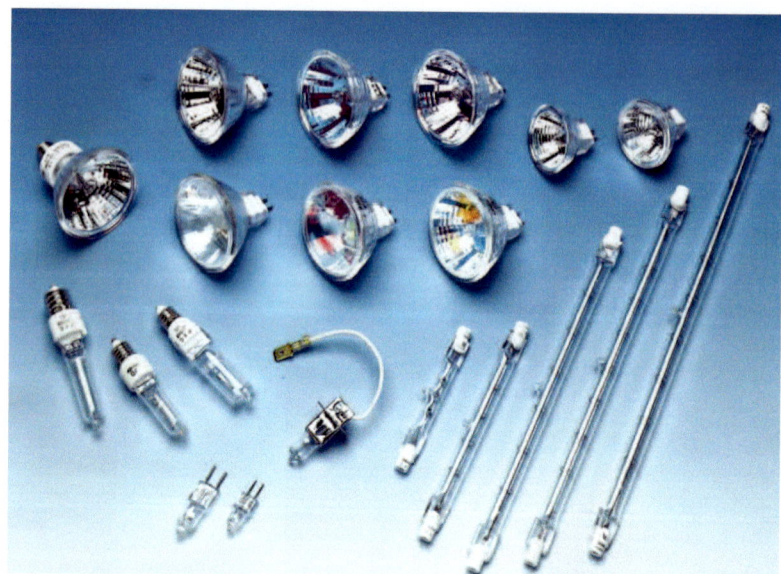

Figure 4.12 A selection of halogen lights.

Source: Hong Kong Lights

Figure 4.13 CFL bulbs are now available in many fittings, sizes and colour temperatures. Over their lifetime they can save up to 40 times the energy of a single incandescent equivalent, being eight times more efficient and lasting five times longer.

Source: Osram

Incandescent

The oldest type of light, deprecated because of its inefficiency, works by passing an electric current through a wire, typically made of tungsten. Halogen lamps are around 30 per cent more efficient than the old-fashioned type. They have good colour-rendering ability but do not last very long, and are also deprecated (see Table 4.1).

Fluorescent

This category includes fluorescent tubes and compact fluorescent tubes (CFLs). These work by passing the electric current through a gas which causes it to glow. The gas needs to warm up before it will work, but with these types of lights it usually doesn't take more than one second. They have good colour-rendering ability. Their lifetime varies from 6,000 hours for compact fluorescents, to 12,000 hours for fluorescent tubes. Long-life tubular versions exist which can last up to 70,000 hours. They can be switched on and off easily and are dimmable.

CFLs now come in many different fittings and styles, including downlights, spotlights, dimmable wall lights, mirrors and so on. Many require 'ballasts'. Most dimmable ballasts require additional wiring, but some are available which do not, for the changing of a centre pendant. For multiple lights, the recall and earth cable is required to carry a permanent live cable and a switch.

LEDs

Solid-state Light Emitting Diodes (LEDs) are the most energy-efficient lights because they have a very long life, typically over 50,000 hours, and use very little energy. They emit a point source of light. They are often integrated into the light fixture, so there is no lamp to replace.

LEDs can last up to 2.5 times longer than CFLs, and 25 times longer than incandescents. This makes them especially ideal for inaccessible places. Their individual light is directed, but modern designs imitate old-fashioned bulbs to achieve omnidirectional effects, and are available in fittings which match those of halogen and incandescent lights. Where a halogen light of this type might consume 40W, the LED equivalent will consume between 1 and 5W. LEDs now come in a full range of colour rendering and temperatures. They can therefore replace most usages for lighting, from street lights and traffic lights to space lighting, fridges and mood lighting.

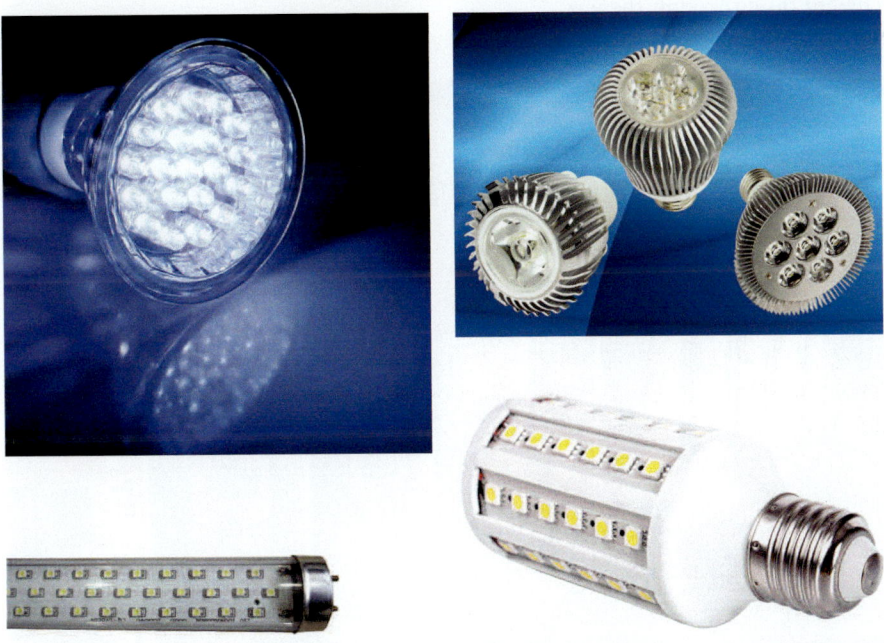

Figures 4.14a, b, c and d LEDs are available in many colours and fittings, including, as may be seen, LED substitutes for fluorescent (in this case T8) and conventional screw fittings, and are eight times more efficient than halogen lights, while lasting 25 times longer.

Source: (a) Crown Copyright; (b) Rayco, used with permission; (c) and (d) GEB Lighting, used with permission

Figure 4.15 6,000K LED street lights.

Source: Author

Gas discharge

Although fluorescent bulbs are also gas discharge, to make a distinction here this category includes sulphur lamps, metal halide and sodium lamps. They work by sending an electrical discharge through an ionised gas, or plasma. They offer long life and high efficiency, but are more complicated to manufacture and therefore expensive, and require auxiliary electronic equipment such as ballasts to control current flow through the gas.

Figure 4.16 LED lighting used in fridges in place of T5 fluorescent tubes.

Source: SunRay Lighting, used with permission

High-pressure sodium lights were traditionally used for warehouses and flood-lighting. Bright orange, they have poor colour rendering and take a long time to achieve full brightness. They are efficient, at 125lm per watt, and last around 20,000 hours. They can be dimmed to a limited extent. Modern equivalents use LEDs.

High-pressure mercury lights produce white light with a bluish tinge, consequently possessing poor colour rendering. They should be phased out. Metal halide and ceramic metal halide lamps offer an efficiency of 80lm per watt and a lifetime of 12,000 hours. Their colour rendering varies from Ra60 to Ra90 and they are available in sizes from 20W to 2kW. They are often used in shops, for exterior lighting and in sports grounds.

Colour rendering and temperature

LED and CFL lights are now available in a wide range of colour temperatures. Colour temperature is a way of describing how cool or warm the colour of lighting is. It is measured in degrees Kelvin, just like normal temperature. This may be found on the packaging. The lower the temperature, the redder, or warmer, the colour. The higher the temperature, the more blue, or cool, the colour. For lamps, Table 4.4 offers a rough guide.

Table 4.4 Colour temperature and perceived colour

Temperature	Perceived colour
2,700K	very warm, yellow white
4,000K	neutral white
6,400K	daylight
8,000K	sky white, almost blue

Colour rendering relates to the way objects appear under a given light source. The measure is called the 'colour-rendering index', or CRI. A low CRI indicates that objects may appear unnatural under the source, while a light with a high CRI rating will allow an object's colours to appear more natural. For lights with a 'warm' colour temperature the reference point is an incandescent light. For lights with a cool colour temperature the reference is daylight. Table 4.5 lists typical colour-rendering index ratings for a variety of lights.

Table 4.5 Typical colour-rendering index ratings for a variety of lights

CRI	Lighting type	Application
22	high-pressure sodium lighting	street lighting
62	common fluorescent tube	office
80–85	compact fluorescent lighting (warm white)	residential
85	premium 4-foot fluorescent tube	retail
80–90	solid-state LED lighting	residential
95	incandescent light bulb	residential

Luminaires

A luminaire is the fixture that holds the lamp. There is now a huge range of types of luminaire, but they all contain five main components: the housing, control gear, lamp holder, lamp and reflector. Reflectors direct light, maximise its usage, and thus reduce the quantity of light needed. Some luminaires may also include a diffuser.

The efficiency of a luminaire is measured by their Light Output Ratio (LOR). The higher this is, the better. An important characteristic for efficiency purposes is the reflectiveness of the material used. Satin chrome reflects only half the light, whereas aluminium coated with silver reflects 90 per cent of the light. Luminaires in offices should be designed to avoid glare onto screens. This type should not be used in retail environments, which have different requirements.

Lighting emissions

As a guide, the annual carbon emissions for a light bulb that is on for four hours per day in a country like the UK where 70 per cent of the electricity is supplied from fossil fuels are as follows:

- 60W incandescent light bulb: 46kg;
- 15W compact fluorescent: 11.5kg (75% less);
- 5W LED 3.8kg (91.7% less).

Lighting controls

Control gear is used to preserve lamp life for gas discharge lighting and can also provide manual or automatic dimming and switching. Modern controls use high-frequency electronic control equipment; this produces more light with less power. However, not all high-frequency lighting can be dimmed.

- Automatic and manual controls should be combined in areas where people need to control their own level of lighting. This means that users should be able to dim, and switch off and on lighting as they require, but that it may also be switched off or dimmed automatically when not needed.
- Dimming gives occupants more control over their light levels and still realises savings.
- Adjustable light-level sensors can automatically turn a light on and off in response to changing amounts of daylight. These may be used outside, such as in car parks or streets, or inside, to maintain an even level of light as the light from outside changes during the day.
- Occupancy sensors can tell whether a space is occupied and control lighting accordingly. They are appropriate for internal or external lighting, and come in several types:

1 Doppler sensors work by sending out high-frequency sound waves and listening for the bounce-back. When it is returned at a different frequency it knows there is a moving object around. It then sends a signal to the dimmable ballast to raise the light levels. When no movement is detected after a certain period of time, the light will return to its original level.

2 Passive infrared (PIR) motion sensors are used for many purposes, e.g. hand dryers and taps, but in the case of lighting can be vulnerable to dust and blocking objects, can be confused by radiators and fires, and have a shorter life span.

3 In some circumstances the use of a manual switch with a timer built into it may be appropriate.

Most control systems need additional connectivity. Nowadays, wireless technology is often cheaper to install than wired connections, as described in Chapter 2. Digital Addressable Lighting Interfaces (DALI) is one global standard used for intelligent lighting management. It is a protocol set out in IEC 60929 and IEC 62386, which are technical standards for network-based systems that control lighting in buildings. A DALI network consists of a controller and one or more lighting devices (e.g. electrical ballasts and dimmers). Each device is assigned a unique static address, allowing it to be remotely controlled. DALI also attempts to reduce the standby parasitic power losses of control equipment.

Upgrading halogen lamps and fluorescent tubes

A range of products incorporating LEDs is now available to replace many display and directional lamps, especially tungsten halogen, but also fluorescent tubes. They combine a light source, power supplier, optics and heat management components. They can be more expensive but last at least ten times longer than the incandescent lamp they frequently replace. It is important to check before purchase if dimming functions are required and that the unit is compatible.

Retrofit kits are available to convert the less efficient non-high-frequency fluorescent light fittings to use T5 fluorescent or LED lamps. (T indicates that the shape of the bulb is tubular, and the last number (y) is the diameter in eighths of

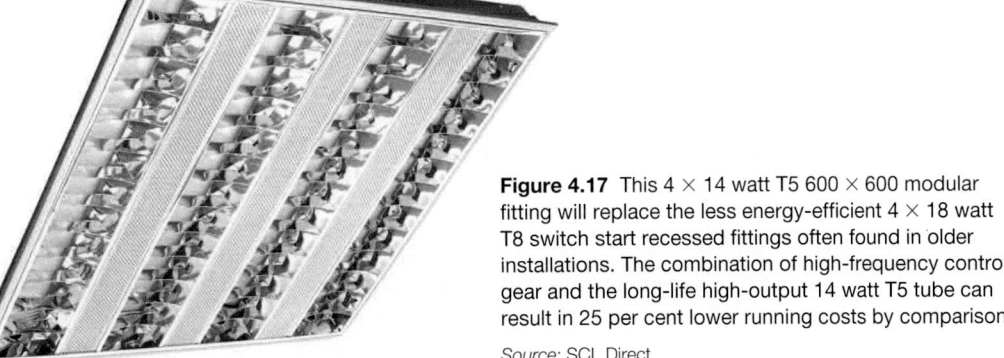

Figure 4.17 This 4 × 14 watt T5 600 × 600 modular fitting will replace the less energy-efficient 4 × 18 watt T8 switch start recessed fittings often found in older installations. The combination of high-frequency control gear and the long-life high-output 14 watt T5 tube can result in 25 per cent lower running costs by comparison.

Source: SCL Direct

an inch.) High-frequency control gear uses around 10 per cent less electricity than the mains frequency equivalent, improves lamp life and eliminates flicker; with mains frequency control gear, for T12 or T8 tubes, lamps often flicker as they are switched on. The LED replacements come in the form of an array, whose fittings match those of the tubes. If the lighting is over ten years old, the fittings should be replaced. This is an opportunity to install lighting controls.

Philips has designed a prototype tube lamp that is twice as efficient as lamps on the market today, yet produces high-quality white light suitable for general use. This tube lighting replacement tube LED (TLED) prototype produces a record 200 lumens per watt of high-quality white light (compared with 100lm/W for fluorescent lighting and just 15lm/W for traditional light bulbs). It is expected to be on the market in 2015 for office and industry applications. In the US alone, for example, fluorescent lights consume around 200 terawatts of electricity per year. If these lights were all replaced with 200lm/W TLEDs, the US would use around 100 terawatts less energy (equivalent to 50 medium-sized power plants), saving more than US$12 billion and preventing around 60 million metric tons of CO_2 from being released into the atmosphere.

Case study: Replacement with LED lights, Wolseley Trade Centre, Birmingham, England

A trade centre retailer installed an energy management system which provided them with data to validate a permanent switch to LED lighting. Following a branch survey, monitoring equipment was fitted, then LED lighting was installed in the mezzanine, trade and office areas and under counters, and linked to the monitoring. It recorded and measured kilowatt-hours usage every 30 seconds, the levels of CO_2 and the cost of each running circuit. The information was available online.

The software quantified energy savings of 64.75 per cent, which cut daily lighting costs by up to 8 per cent and reduced the trade centre's carbon dioxide emissions by the same amount (Table 4.6).

Table 4.6 Savings made by switching lights to LED

Area	Before installation			After installation		
	kWh	CO2 (kg)	Cost (£)	kWh	CO2 (kg)	Cost (£)
Mezzanine	9.73	5.22	1.04	1.97	1.06	0.2
Office	7.12	3.83	0.71	2.83	1.52	0.27
Trade counter	16.54	8.88	1.65	7.83	4.18	0.78
Trade area	10.71	5.75	1.07	5.13	2.76	0.51

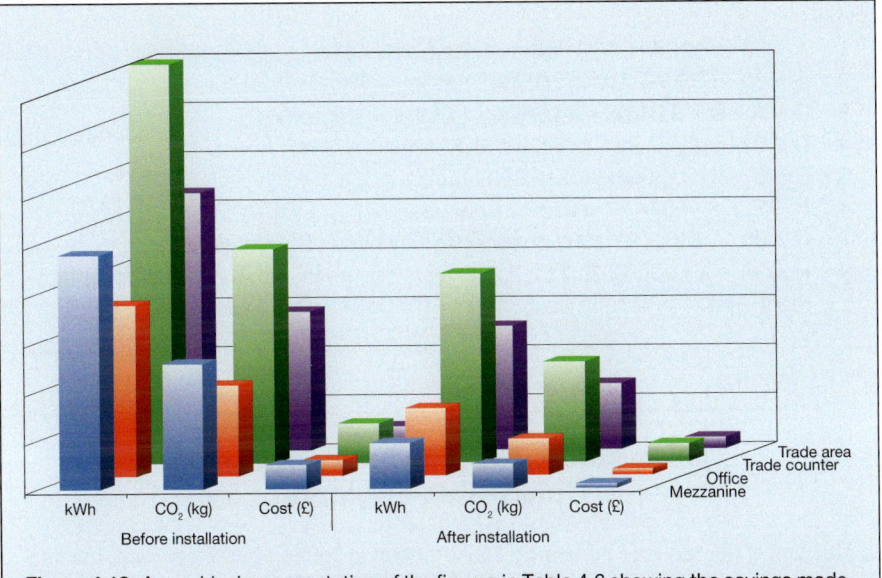

Figure 4.18 A graphical representation of the figures in Table 4.6 showing the savings made.

Source: Author

Estimating the payback period for lighting replacement

1 Find the total power use by adding the power rating of all fluorescent lamps to the power consumption of the control gear (A).
2 Do the same for the new fluorescent or LED lamps and control gear (B).
3 Subtract B from A. Convert to kilowatts (kW) (C).
4 Estimate the annual operating hours of the system (D).
5 Multiply C by D to give the annual electricity usage (E).
6 Multiply E by your current electricity cost (F).
7 Obtain an estimate of capital cost and time cost of the replacement equipment (G).
8 Divide G by F to find the simple payback period in years (not accounting for inflation) (H).

Example

A room containing 30 100W T12 fluorescent lamps, not using high-frequency control gear, is to be replaced with T5 lamps using retrofit kits costing £40 each including installation. The lights are typically on from 9 a.m. to 5 p.m. Monday to Friday. Electricity costs 13p/kWh.

1   A = 30 × 100W + 13W [control gear] = 3,013W.
2   B = 30 × 35W + 4W = 1,054W.
3   C = A − B = 3,013W − 1,054W = 1,959W = 1,959kW.
4   D = 8 hours per day × 5 days × 52 weeks = 2,080 hours.
5   E = C × D = 1,959kW × 2,080 hours = 4,074.72kWh per year.
6   F = 13p × 4,074.72 = £529.71 per year.
7   G = 30 × £40 = £1,200 (labour cost may need to be added).
8   H = G/F = £1,200/£529.71 = 2,265 years of operation to pay back the replacement cost.

Case study: The University of Essex, Colchester, England

This university reduced its energy consumption in lighting by 77 per cent, making savings of £10,311 a year. With the help of energy products and services distributor Rexel, and at a cost of £1 million, it replaced the external walkways of the entire university campus. The lighting levels had to be high enough to create an environment in which students would feel secure when moving around the campus.

A total of 150W HQL tubular lamps were replaced with 197 Kingfisher LED-IN1BC4.7. The fall in energy consumption was from 133,509kWh to 30,396kWh per annum (a saving of 103,113kWh). The university's carbon footprint was also reduced by 56 tonnes of CO_2 per annum.

Before the new lighting was approved, samples from various manufacturers were installed in a small area of the campus for testing and comparison purposes. On this basis, a cost-effective product was selected, with a price that enabled a higher number of the street lights to be retrofitted.

Brian Smithers, strategic development director for Rexel Northern European Zone, comments: 'Lighting can be the quickest win for organisations wanting to cut bills and reduce their energy consumption. As more organisations start to think about their energy costs, lighting is often the first area they address, as installation is quick and painless and subsequent results speak for themselves.'

Disposal of old lights

Old lighting that is being replaced should be disposed of thoughtfully. In Europe, under the Waste Electrical and Electronic Equipment (WEEE) Regulations, the producer of electrical equipment has a responsibility to finance the recycling of their products when they reach end of life, which they do by joining a compliance scheme, and this manages the process on their behalf. Businesses can save money and benefit the environment by contacting the producer compliance scheme which represents the producer of the equipment due for recycling.

5
Heating and cooling

This chapter looks at methods of heating and cooling which use power, including solar power. It also looks briefly at methods of heat recovery from equipment. This topic, together with minimising heating and cooling costs in buildings and using passive solar heat gains, is covered in much more detail in our companion volume, *Energy Management in Buildings*. Chapter 6 examines complete HVAC (heating, ventilation and air-conditioning) systems.

Heat recovery

With the objective in mind of reducing heat demand and taking advantage of passive gains before calculating the remaining energy required to heat and cool a building, opportunities to recover heat from other processes should be examined first.

Ventilation heat recovery

In a typical ventilation system with heat recovery, the heat from stale air being expelled from a building is transferred, using a heat exchanger, to incoming fresh air, by passing the incoming air over a series of pipes which contain the hot outgoing exhaust air. Buildings need to have a high standard of airtightness. Efficiencies of these systems can vary from 55 to 85 per cent.

Other types of heat recovery

Free heat may be recovered from many processes that take place within commercial and industrial premises, where it would otherwise be lost. The use of this heat to supplement or even replace fuel used specifically for heating and cooling can provide rapid returns on investment by removing the need to purchase fuel or electricity. Emissions savings achieved thereby should also be quantified so that recognition can be gained.

Boilers, examined below, provide a common example of heat recovery. Other sources of 'waste' heat can be air compressors, refrigeration, hot liquid effluent, power generation, process plant cooling systems, and high-temperature exhaust gas streams from furnaces, ovens, kilns and driers. Some of these sources are examined in further chapters.

Reclaimed heat may be used for any purpose, even for cooling, by driving a heat engine. Besides heating incoming fresh air for ventilation, the most common targets for use of the waste heat are the preheating of combustion air for boilers, ovens and furnaces; preheating boiler feed water for hot water systems; drying; preheating for other industrial processes; and space heating. Low-grade heat can be concentrated into higher-grade heat.

Choosing a fuel for additional heating

Electricity, gas, solar thermal, heat pumps (air, ground and water sourced), oil and biomass are the choices for heating. Electricity may be locally produced and renewable or grid sourced. Oil may be fossil fuel-based or sourced more sustainably from biofuels and recycled oil or cooking oil. Equally, gas may come from the gas grid, be LPG (liquid petroleum gas), or may be renewably produced biomethane from anaerobic digestion plants or landfill gas (Table 5.1).

In terms of overall climate change impacts, the favoured options would be as follows:

- Sustainably-sourced biomass (but see p185) gas from renewable sources (biomethane, from anaerobic digestion), and oil from renewable sources (biodiesel) or recycled oil. All of these would feed combined heat and power cogeneration (CHP), meaning that the same plant would produce both heat and electricity, the most efficient type of heating plant (up to 95 per cent efficient). Most successful plants of this nature feed the hot water through a heat main throughout the site; a district heating system. The electricity produced may be used or sold locally, or exported to the grid.
- Directly supplied, locally produced electricity, whether from CHP or renewable sources such as hydroelectricity, wind or photovoltaic solar, is more efficient than grid electricity because transmission costs over the grid may lose up to 10 per cent of the original power and result in higher supply costs.
- A solution involving a combination of heat pump and/or solar thermal is also an option, with the heat output being stored in water tanks. It is even more desirable if the electricity powering the pumps is renewably sourced.
- The next favoured options would be energy-from-waste CHP, then CHP using gas, either LPG or from the grid. Gas has half the carbon impact of coal and oil (around $400kgCO_2/MWh$). For smaller buildings a condensing boiler is an efficient option.
- In general, using electricity for heating is not recommended, if coming from the grid, in areas where the grid has a high proportion of fossil fuel-burning power stations, since it is better to use the fuel burnt in these power stations directly for heating water than convert it to electricity first and then back to heating.
- Finally, and to be avoided if at all possible, are oil- or coal-fired heating, because of their impact on climate change.

The remainder of this chapter discusses sustainable heating.

Table 5.1 Classification of heating fuels

	Gas	**Oil**	**Biomass**	**Solar**	**Heat pumps**	**Electricity**
Renewable	anaerobic digestion	used vegetable oil	Wood pellets	solar thermal panels (flat plate collectors)	ground source	wind
	landfill gas	used mineral oil	logs	evacuated tubes	air source	solar photovoltaic
		biodiesel*	woodchips	concentrated solar thermal	water source	hydroelectric
			waste/garbage		geothermal	fuel cell
Nonrenewable	LPG	mineral oil	coal			coal
	gas mains					gas

Note * Biodiesel may be produced from soybeans, canola, oil palm or sunflowers (not so sustainable), animal fats and used cooking oil (the most sustainable). Some is produced from commercially available methanol, which is often made from the methane that exudes from oil and gas wells, and burning it, or a methyl ester derived from it, continues to exacerbate climate change.

Reused oil as a fuel

The case study below from the Irish Navy is an example of the use of recycled oil as a fuel for heating. This is ideal if an organisation produces a large and reliable supply of used oil which would otherwise be thrown away. There may also be a local source of used vegetable oil, for example, from catering businesses, or of biodiesel, which is chemically almost identical. All of these types of oil may be burnt in a CHP plant. Existing oil-burning boilers may be converted to consume such fuels.

Case study: Haulbowline Naval Base, Ireland

Figure 5.1 Haulbowline Naval Base, Ireland.

Source: Irish Department of Energy

A waste oil boiler was incorporated into a building upgrade for mixed-use workshops, offices, toilets and canteen at Haulbowline Naval Base, Ireland. M&E consultants found that every year naval ships produce 70,000 litres of usable waste engine oil, while ships' kitchens and cookhouses produce 12,500 litres of waste vegetable oil. They were instructed to install a boiler fired by this waste oil to provide space heating and hot water to the facility.

The volume of oil required to fuel the boiler was estimated to be 25,000 to 35,000 litres per annum. If diesel was used, just 20,000 litres would be required, due to the higher energy content in diesel compared to waste oil. The additional capital cost of the boiler was €7,300. Design and consultancy fees were €2,500. Additional valves, etc. cost €1,800. The total extra cost for installation of the waste oil boiler was €13,166, including 13.5 per cent VAT. This saved 20,000 litres of diesel at €0.6 per litre, or €12,000 per annum. The payback period was 13 months.

The yearly additional running cost due to filters, extra maintenance and so on was €1,400 including VAT, so after 13 months there have been savings of €10,600 per annum. As oil prices increase, savings increase. If used as the only oil source, vegetable oil would make the boiler system more low carbon. Local restaurants may provide such oil free of charge. This project has potential for expansion to other sites in the defence forces.

Boilers or furnaces

Figure 5.2
A Viessmann Vertomat condensing boiler installed in a multi-unit apartment building in White Rock, British Columbia, Canada.

Source: Wikimedia Commons (Dunnd74)

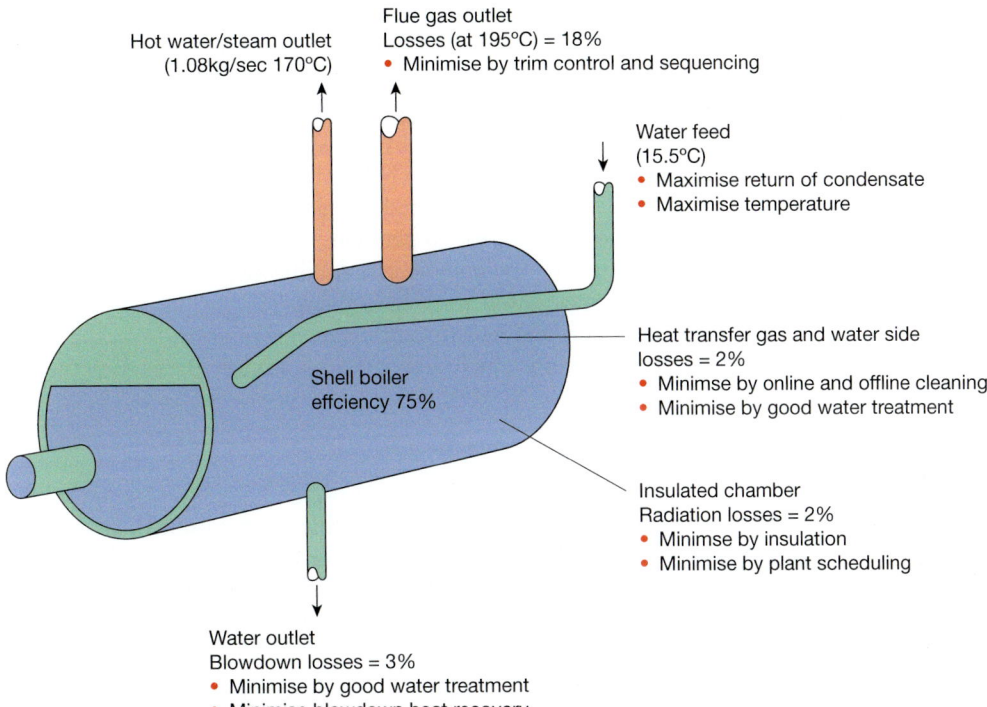

Hot water/steam outlet
(1.08kg/sec 170°C)

Flue gas outlet
Losses (at 195°C) = 18%
• Minimise by trim control and sequencing

Water feed
(15.5°C)
• Maximise return of condensate
• Maximise temperature

Heat transfer gas and water side
losses = 2%
• Minimse by online and offline cleaning
• Minimise by good water treatment

Shell boiler
effciency 75%

Insulated chamber
Radiation losses = 2%
• Minimse by insulation
• Minimise by plant scheduling

Water outlet
Blowdown losses = 3%
• Minimise by good water treatment
• Minimise blowdown heat recovery

Boilers (UK) or furnaces (USA) employ one or more hot water tanks for storage and contain a combustion chamber through which pass the hot gases created by burning gas or oil. With efficient condensing boilers, this is surrounded by a heat exchanger which transfers heat from the gases into the water inside. The cooler, heavier flue gases are extracted from the boiler by a fan. Then, a second heat exchanger removes even more of the heat from the flue gases. It pre-warms water coming back into the boiler from a heating system, so less gas is required to heat the water up. By using this otherwise wasted heat, condensing boilers use around 90 per cent of the heat they generate, compared to old boilers without this facility.

Since the waste gas loses some of its heat, it cools down into an acidic water called condensate and water vapour, giving the boiler its name. The cooler, heavier flue gases are blown out of the boiler by a fan. Manufacturers claim that up to 98 per cent thermal efficiency can be achieved; but a field trial conducted by the UK's Energy Saving Trust[1] found an average efficiency of 85.3 per cent. Because the water they heat is stored in a tank, they can be used in combination with other heat sources such as a heat pump, solar water panels or electric immersion. Always choose an A-rated boiler.

Boilers should be sized on the basis of hot water requirements. This will also reduce water use, a subject covered in relation to heating systems in much greater detail in Chapter 12. There is a trend towards increasing the electricity use of these boilers by fans, pumps and control systems. Average combination boilers use around 30 per cent more electricity to supply 10,000kWh of heat than regular boilers, and around 50 per cent more electricity to supply 20,000kWh of heat. This consideration should be a factor in the choice of heating supply.

Figure 5.3 The sources of losses from shell boilers (the most common type). Most boilers of this type can be made to run at 80 per cent efficiency.

Source: The Carbon Trust

Larger boilers can have their efficiency improved in three ways:

1   Improving their combustion efficiency, by checking the operation of the burner, its controls, and removing unburnt fuel, soot and ash; they should be tuned to the correct temperature by adjusting the fuel-to-air ratio on a regular basis. It should also be checked whether the burner is firing at a rate too high for the boiler to which it is fitted. Together, these measures can improve efficiency by between 5 and 30 per cent.
2   Improving transfer efficiency, by removing deposits on the heat transfer surfaces, using specialised water treatment.
3   Minimising boiler heat loss through heat reclamation from the flue gas and by insulating the boiler, including checking for any damage to the insulation and repairing it.

Heat recovery from boilers

The operational efficiency of the boiler is defined as the useful heat output divided by the fuel input. Many boilers lose heat, for example, through the flue (about 18 per cent) and in heat transfer from the water or gas. Boiler flue economisers, if not already present, can be retrofitted to most steam and high-temperature hot

Figure 5.4 A schematic diagram of a dual fuel condensing economiser. Heat from the waste flue gases is transferred to the incoming water.

Source: Public domain (shared by haj90599)

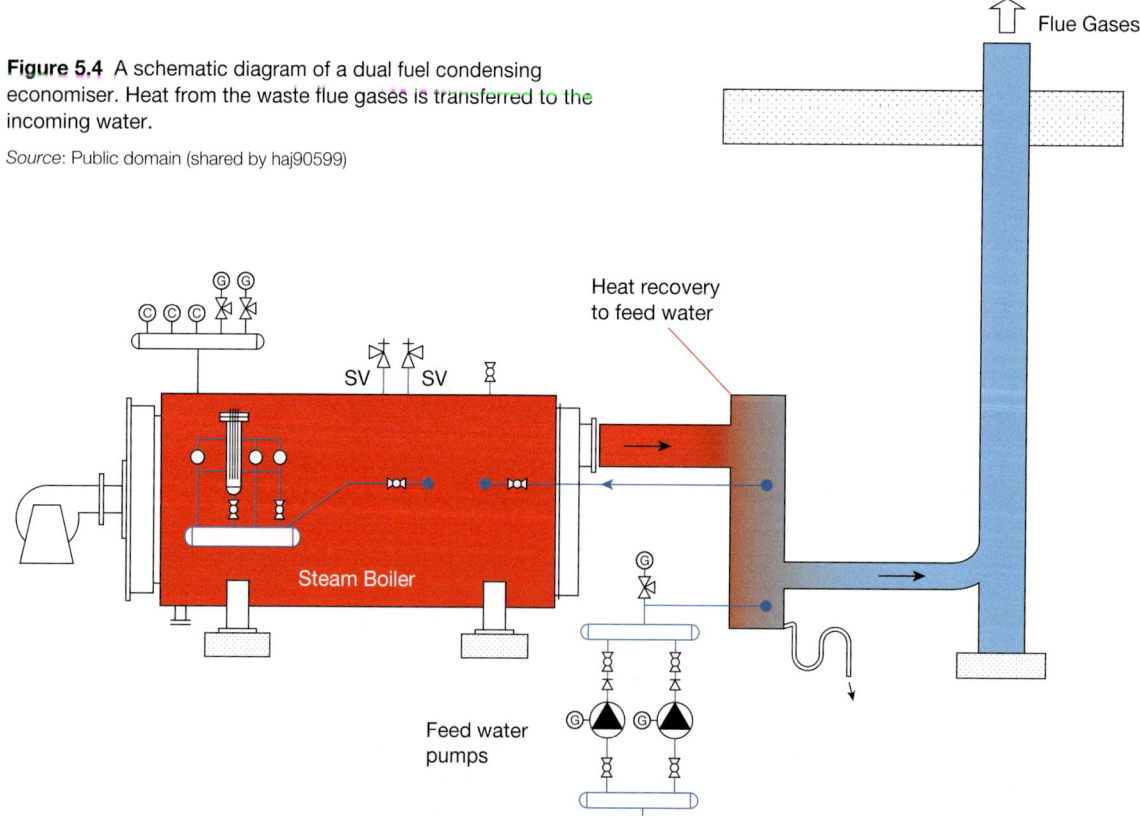

water boiler flues and often to non-condensing boilers. Exact savings depend on the type of boiler to which they are fitted.

The heat recaptured may be used to preheat boiler feed water or stored in a tank to provide hot water for other purposes. The economiser must be sized correctly for the flue, and the flue gases must not be condensed to liquid when they arrive at the economiser. Feed water must not boil in the exchanger. If the economiser is designed to condense flue gases, the water returning from the heating circuit must be cool enough to get the benefit of the additional heat, i.e. below 50°C (122°F) for hot water boilers and below 90°C (194°F) for steam boilers.

Preheating the combustion air feeding into the burner to the same temperature as the boiler also improves efficiency by 1–2 per cent. The source for this heat may be the heat remaining in the flue gases, a higher-temperature air drawn from the top of the boiler house, or heat recovered by drawing air over or through the boiler casing. In the first case, outside air is drawn through the boiler flue economiser and ducted to the burner air input. In the second case, air is drawn with a fan from the ceiling level. In a typical office of about 250 people this cheap measure has a rough payback of five years. For a steam boiler, blowdown heat recovery has a similar payback period. The heat is used to heat the feed water. It is unlikely to be cost-effective on boilers below 1,000kW.

Heat recovery steam generators (HRSGs)

These are boilers or furnaces that recover exhaust heat from gas turbines in order to generate steam for the steam turbines. Their market is expected to grow substantially, as part of a long-term goal to meet carbon emissions targets by large companies. They are used in combined cycle gas turbine (CCGT) plants.

CHP or cogeneration

Combined heat and power, or cogeneration, removes the need to have separate electricity generators and boilers by using heat recovery to reclaim most of the 60 per cent or so of the energy output that is otherwise lost as heat. Usually, an organisation wishing to install such a system will commission an external supplier who can offer a professional partnership arrangement for developing new energy schemes and can provide all or part of a system. The service arrangement covers delivery of the power and heat, which means that the plant is delivered as a turnkey operation, and the service company is responsible for all maintenance. CHP plants may also be used to supply cooling. This is known as tri-generation, or CCHP. This is ideal for sites which require a large amount of air conditioning. It displaces the need for separate air conditioning, reducing overall carbon dioxide emissions by as much as 29 per cent. Substantial savings are possible.

Industrial CHP systems range in scale from a few megawatts to the size of a normal power station. Large installations with their own power generation requirements find multiple advantages in installing CHP. The surplus heat may be used for a variety of purposes around a site. For example, British Sugar has a sugar beet factory that is among the most efficient in Europe, located at Wissington, Norwich. Its 70MW gas-fired CHP plant (pictured in Figure 5.6) meets all of the steam and electricity demands of the sugar production line, but

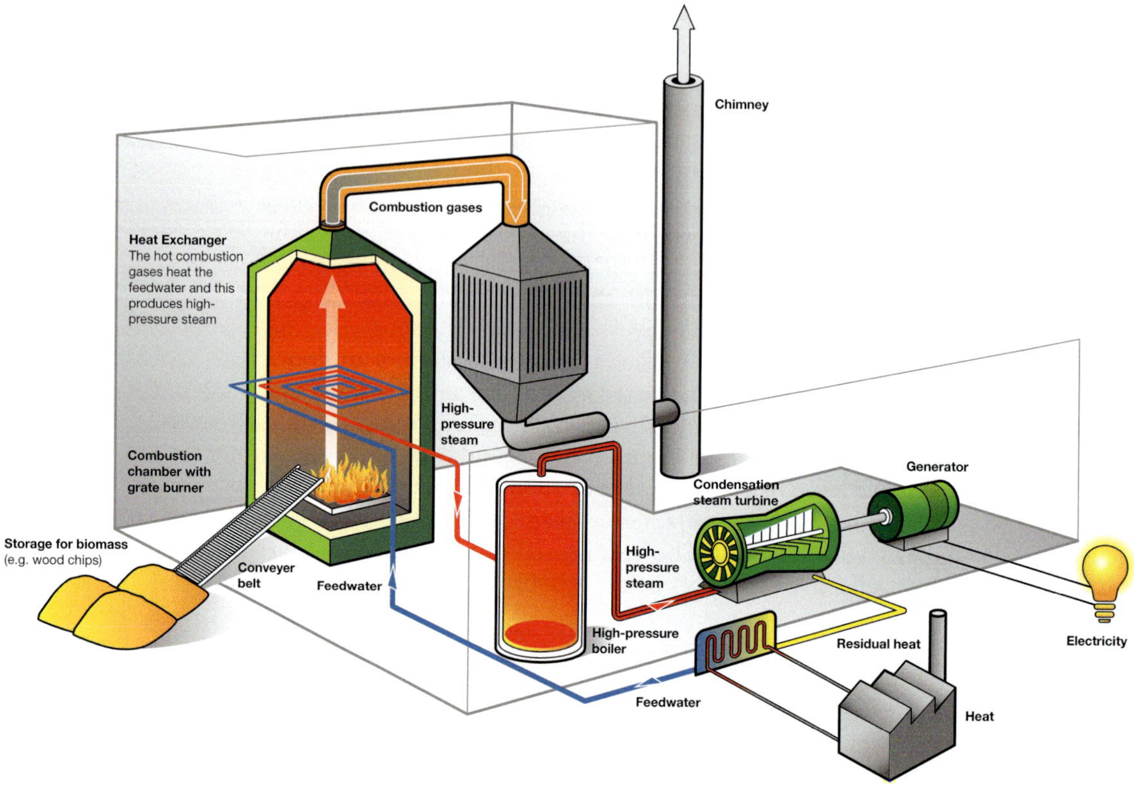

Figure 5.5 Schematic diagram of how a combined heat and power, or cogeneration, plant works. In this case, the plant runs on biomass (woodchips or pellets).

Source: Wiki Commons: Jonathan O'Reilly

Figure 5.6 British Sugar's CHP plant.

Source: British Sugar

also exports 50MW of electricity to the local network. The flue gases, which would normally be expelled into the environment, are diverted to an adjacent glasshouse, providing heating and carbon dioxide that is essential to promote plant growth. The greenhouse covers 11ha and produces 70 million tomatoes per year, a real example of thinking holistically.

District heating CHP systems are becoming more widespread. In Sheffield, England, over 140 buildings are connected to a scheme, including hospitals, shops, offices, the theatre, to universities and the City Hall. They are connected by over 44km of underground pipes generated from an energy-from-waste plant that produces 60MW of thermal power and up to 90MW of electrical power. It saves around 21,000 tonnes of carbon dioxide per year compared to electricity from the grid.

Mini-CHP systems are also available. These meet the requirements of large and medium-sized stand-alone buildings. For example, supermarket chain Sainsbury's has two stand-alone packaged gas-fired CHP units at a store in West London, each of them sized at 210kW. These have reduced energy bills by £20,000 per year, and CO_2 emissions by almost 2,000 tonnes over the same period.

Figure 5.7 A series of mini-CHP units. These have a small commercial scale application and are rated with an electrical output of 4–13kW and a thermal output of 17–29kW. Units sized much smaller than this are not usually economic to run unless they are on constantly.

Source: EC Power

Biomass boilers

Biomass is not entirely carbon neutral, depending on how far away the fuel is sourced, and whether it is in the form of pellets, which are energy-intensive to produce, basic timber or woodchips. Therefore the fuel should be available, reliable, locally sourced, accessible and appropriate. As an example, a 150kW

Figure 5.8 This biomass woodchip boiler is fed from a large hopper, or storage container, above ground. Note the well-insulated pipes.

Source: Author

woodchip boiler will use around 400 tonnes of seasoned woodchip a year. This would require around 5,000 to 7,000 acres of nearby woodland from which the pollarded or coppiced wood could be taken and a service agreement set up to do so. The fuel's moisture content also affects is calorific value, so it must be dried and seasoned. Every attempt should be made to size the plant correctly, since if it is over-specified it will be running inefficiently if it is asked to output at a lower temperature than its optimum operating temperature. Biomass boilers require more supervision than a gas boiler.

Anaerobic digestion (AD)

Anaerobic digestion (AD) provides biogas that may be used directly to produce heat and electricity in an on-site CHP plant, or even used as a fuel in some vehicles. Surplus gas may also be sold to the gas grid after being treated, and the electricity from the CHP plant may be exported or used locally. The feedstock is any organic material: food, agricultural and gardening waste. A guaranteed year-round supply is necessary. Therefore, the owner must either have this available themselves (they may be a food producer or retailer, or agricultural concern) or source additional material from similar nearby companies, which may pay to have their material taken away, since they would normally pay disposal costs. AD also provides a fertiliser which may be used directly on the ground or sold.

Figure 5.9 Schematic diagram of an anaerobic digestion plant. The composting of the organic material (feedstock) occurs in the process cylinder.

Source: Author

Energy from waste

The incineration of waste is less efficient than anaerobic digestion, and should therefore be reserved for non-organic materials. It entails burning the dried waste

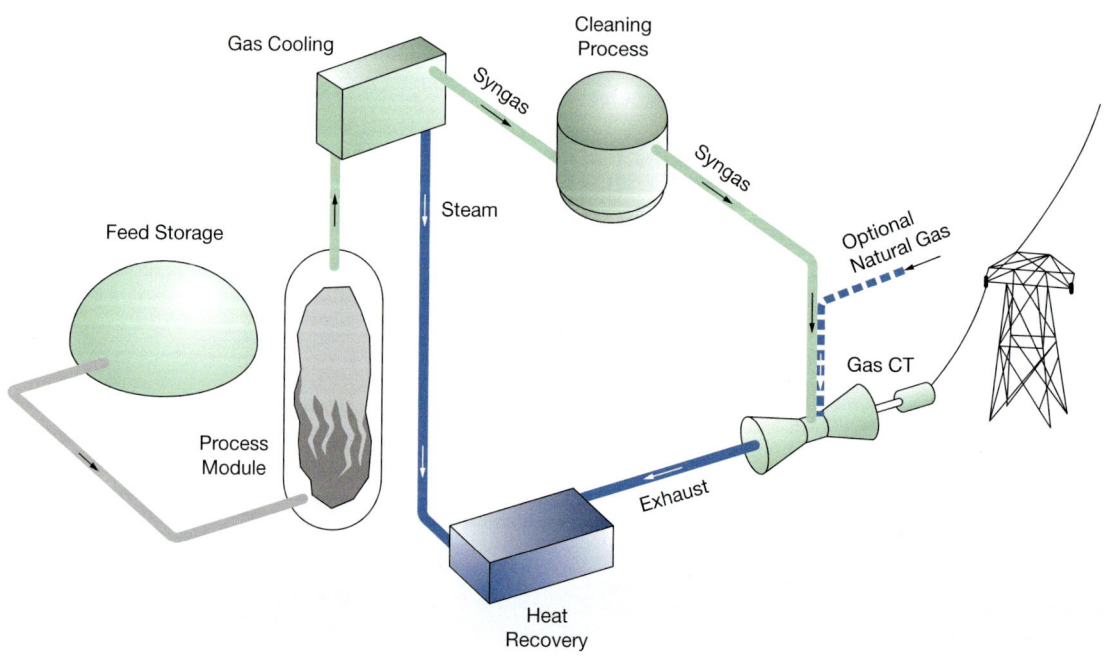

to drive a steam-powered CHP plant, which produces electricity and heat. Energy from waste, like AD, requires a constant input of suitable combustible material. Additional challenges involve drying the waste and ensuring that no pollutants are released into the atmosphere, in order to meet the requirements of local pollution regulations and the environmental monitoring regime.

Opponents of this technology argue that it discourages the recycling and reuse of waste, and that even legislation-compliant plants produce some air pollution. An incinerator may be useful if a facility produces a constant amount of a particular waste that cannot be disposed of in more sustainable ways and would otherwise go to landfill; at least the energy content of the waste material will be recovered.

Heat pumps

Heat pumps draw the free heat from the sun that is stored in the ground or the air, to concentrate it in a much smaller area, i.e. a building, using a pump. Heat pumps are an efficient technology because the ratio of the energy input in the form of electricity to power the pump to the energy output in the form of heat is designed to be at least 3:1. This ratio is called the coefficient of performance (COP), and the higher it is, the more efficient the unit. In this case, for each unit of electricity put in, three or more units are produced.

The exact efficiency in practice depends on the difference in temperature between the target and the source at any one time. If the target temperature is underfloor heating at 18°C (65°F) and the source is 2°C (36°F), then less energy is required to concentrate and pump the heat than if the source is colder (as in air-source heat pumps in freezing weather) or if the target is hotter (as in a radiator-based heating system where the target temperature may be 60°C (140°F)). This is why ground- and water-source heat pumps are generally more efficient than air source; in winter, the temperature of the air outside is generally much lower than that of the ground three metres (3.28 yards) below the surface.

Heat pumps can be confused with geothermal energy. Strictly speaking, geothermal energy is derived from hot rocks that can be relatively close to the Earth's surface in some parts of the world. It requires a very deep borehole to access them: 0.5 to 1 kilometre. Water is pumped down, heated up by the hot rocks and pumped back up again. It may be used directly in a district heating system. Ground-source heat pumps may also have boreholes, as shown in Figure 5.10, but they do not use geothermal energy. These boreholes may be up to 100m deep. Vertical boreholes are more expensive to create than horizontal trenches (typically 2 to 4 meters (2 to 4 yards) deep see Figure 5.11) and contain a loop, but they are an option if space is not available to dig a trench.

Heat pump manufacturers' estimates of their COPs should be treated with caution, because real operating conditions will not reflect the test conditions. The British Standard used to test and quote for most packaged heat pumps is BS EN 14511. This, for example, specifies test conditions of 7°C (45°F) outdoor (source) air temperature for air-source heat pumps and a return and flow temperature of 40°C (104°F) and 45°C (113°F) respectively. It is quite possible for air-source heat pumps to use more energy than the heat source they are replacing if specified and installed incorrectly.

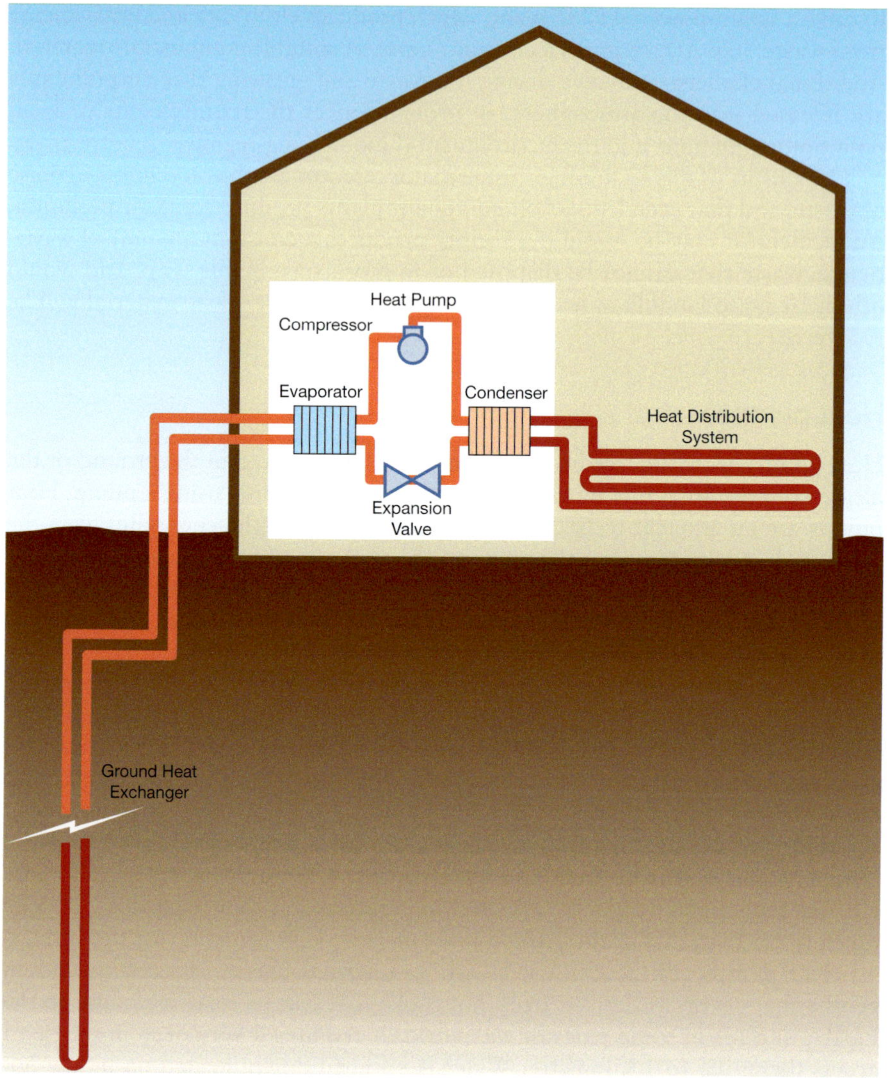

Figure 5.10
Schematic diagram of a vertical loop ground-source heat pump. An air-source system would be similar, but without the underground loop; instead, the left-hand heat exchanger would take heat from the outside air drawn into the unit.

Source: Energy Efficiency Best Practice Programme Factsheet

Figure 5.11
A horizontal trench with a collection coil containing the water for a ground-source heat pump.

Source: John Cantor

Heat pumps can transfer their heat to air or water. If to water, this is to the underfloor heating system. If to air, a condenser inside heats the air at the point where it is supplied to the building. Filtered, pre-warmed air is directed into the building from vents by a ground-floor wall. One advantage of air-destination heat pumps over the water-destination variety is that air into which the heat is passed typically has a lower temperature (called the sink temperature) than that of water. This results in a higher COP and increased heat output.

Solar water heating

Solar water heating (solar thermal) works well in areas with east-south-west-facing roof space (in the northern hemisphere). It is often combined with other sources of heating, such as biomass and heat pumps. The storage tanks permit multiple inputs from the different sources of heat which then transfer their heat from closed loops in their respective systems.

On average, throughout the year, around 60 per cent of water heating requirements can be met this way in medium latitudes and 40 per cent in higher latitudes (depending on collector area and system efficiency). Closer to the equator, almost all water can realistically be solar heated throughout the year. As long as the temperature in the collector is higher than that of the incoming cold water (usually about 10°C (50°F), then a solar water-heating system will save energy.

Sealed, indirect solar systems form the majority of industrial installations. They typically use the more efficient evacuated tube collectors, which consist of long glass tubes containing a metal strip inside a vacuum that absorbs the heat and transfers it to a liquid at the manifold end. These units are modular, with the advantage that they can be added to and installed quickly.

Another type of solar collector, frequently but not always ground based, consists of rows of parabolic reflectors which, tracking it throughout the day, concentrate the sun's light on to a tube positioned at the focal point containing a similar liquid. Again, this is transported to manifolds and then to the heating

Figure 5.12 A simple solar water-heating system. The tank permits inputs both from these solar collectors and another auxiliary boiler, which could be gas, biomass, or even a loop from a heat pump. The outputs are for space heating and hot water.

Source: Author

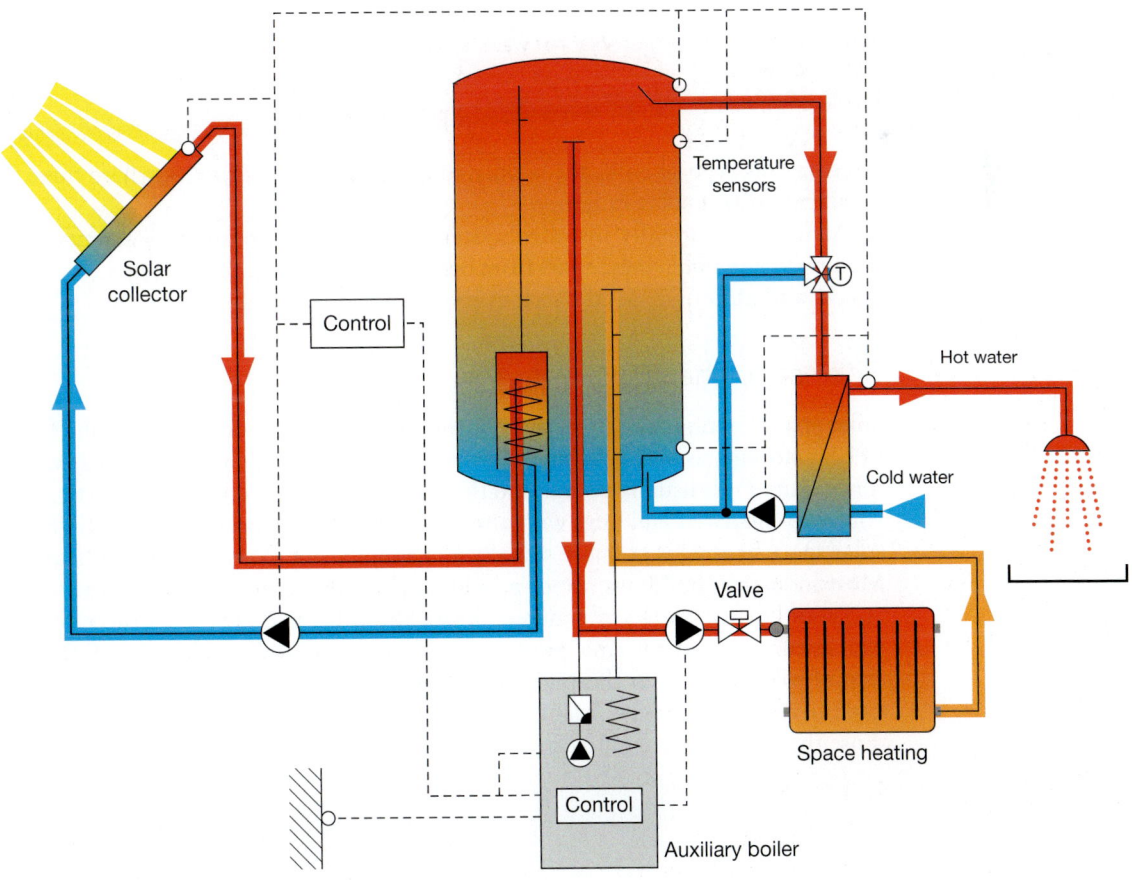

Figure 5.13
Evacuated tube
collectors in
Wezembeek, Belgium.
They produce around
6.6 megawatt-hours
(MWh) of heat energy
per year.

Source: IEA-SHC
(International Energy
Agency-Solar Heating and
Cooling programme)

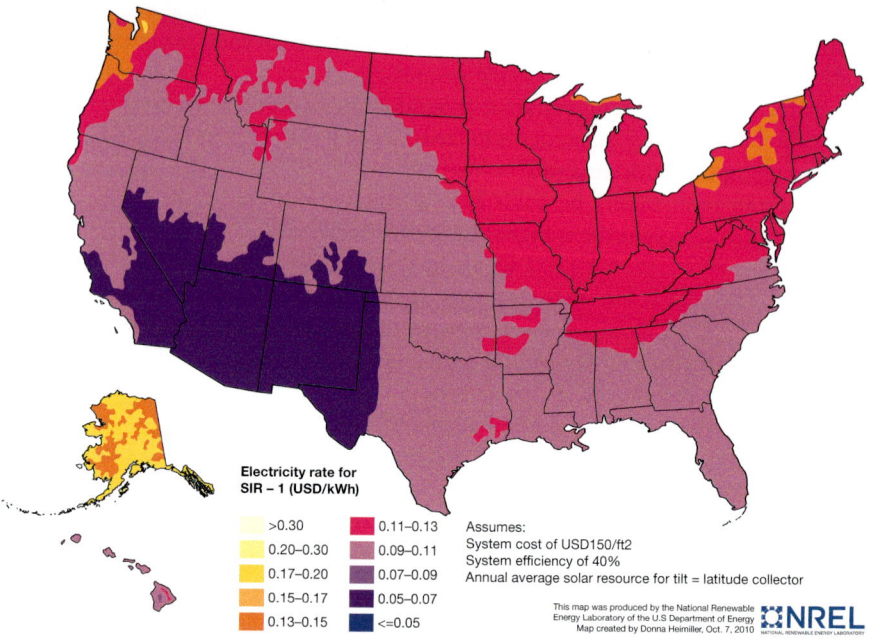

Electricity rate for
SIR – 1 (USD/kWh)

>0.30	0.11–0.13	Assumes:
0.20–0.30	0.09–0.11	System cost of USD150/ft2
0.17–0.20	0.07–0.09	System efficiency of 40%
0.15–0.17	0.05–0.07	Annual average solar resource for tilt = latitude collector
0.13–0.15	<=0.05	

This map was produced by the National Renewable
Energy Laboratory of the U.S Department of Energy
Map created by Donna Heimiller, Oct. 7, 2010

NREL
NATIONAL RENEWABLE ENERGY LABORATORY

Figure 5.14 The viability of solar hot water systems compared to electric water heaters in the United States.

Source: NTREL 2011

system. This type of collector is only suitable for hotter climates where there is significant direct solar radiation.

Most professionally installed systems come with a ten-year warranty and require little maintenance, apart from occasionally cleaning the collectors, a yearly check, and a more detailed check every three to five years.

Solar space heating

Solar space heating uses some kind of solar collector and, optionally, some form of heat storage, as well as a heat distribution system. The challenge is that when it is required it is usually winter and there is not so much sunshine. Nevertheless, with well-insulated buildings with an HVAC system with heat recovery (MVHR), solar space heating is possible, usually as a supplement to other forms of heating. Most industrial buildings are of a lightweight metal construction for the walls and roof but with a heavy concrete floor. The walls and roof must therefore be well insulated. Solar thermal plants in industry currently form less than 0.05 per cent of worldwide solar thermal capacity. Only some of this is for space heating, and the rest is for process heat. Solar space heating systems are divided between those that use warmed air as a delivery medium and those that use water.

Air-based systems

The simplest system heats air drawn directly into a space by natural ventilation or forced ventilation, using fans. They are really only appropriate for climates with long, cold winters that have many sunny days. The solar heating air, heated using a glazed system with a large collector area, is allowed to enter the occupied space slowly and continuously from several different points. The most efficient operating temperature is around 32°C (90°F).

For large warehouse-type buildings, collectors come in the form of cladding on the sun-facing exteriors. One design features glazing over a black metal plate that draws air from ground level within the building and heats it as it is drawn up by convection or a fan between the two layers, then directing it back into the building at ceiling level.

A further design, the transpired solar wall, consists of metal sheets perforated with thousands of tiny capillaries, each absorbing the sun's heat. The solar wall is positioned 8 inches (20cm) away from the inner wall. Incoming air is again drawn through the holes, up the space in between where it warms up, and through ducts into the building where it is discharged around the ground level. It rises through the building using the stack effect. Alternatively, it may be drawn into a rooftop MVHR system. Ducting transmits the heated air around the building.

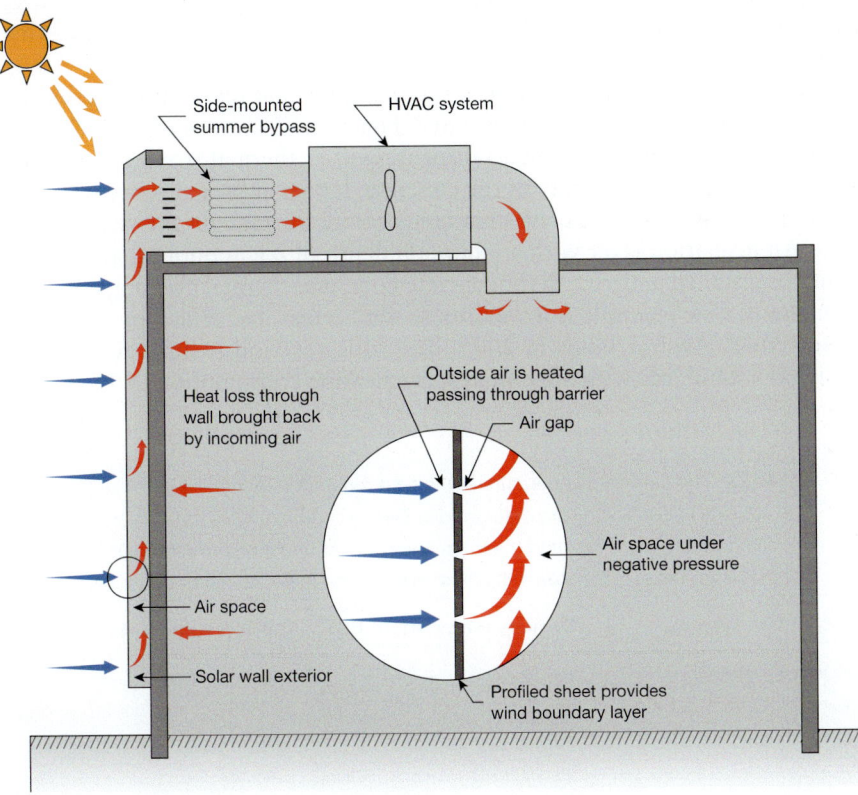

Figure 5.15
Cross section of the transpired solar wall principle integrated with a HVAC system.

Source: Conserval Engineering

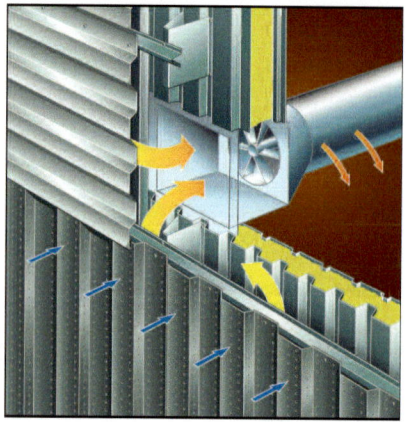

Figure 5.16 Cross section of the transpired solar wall principle.

Source: Tata Steel

More conventional collectors can also be mounted on a roof, in which case they are connected into an MVHR system. This preheats the fresh air that is input into the ventilation. If the incoming air is warm, the recovered heat from the outgoing air is not needed and the heat exchanger is bypassed. At night, and on colder days, the incoming air from the solar panel has the recovered heat transferred into it. Because the air's moisture content is normally low when the sun shines, these systems help to dehumidify the internal climate.

Water-based systems

Water-based solar space heating systems are commonly hybrid or combi systems, preheating the water so that another heat source has less work to do, as described above. Even in Sweden or Canada, it is possible to halve primary energy use with such a system. The challenge is to size it correctly, so that the panel collector area and the storage tank volume match up at the most efficient level. This depends on local weather and latitude, the available roof area, and the pattern of use of hot water and building occupancy. The larger the solar collector area, the faster the storage tank will heat up; but too large and collected heat will be wasted.

Space heating from solar water and from heat pumps is at its most efficient when delivering heat at a low temperature to underfloor heating systems, as may be seen from Table 5.2.

The concrete in the floor acts as a thermal store and maintains an even temperature, provided that it is well insulated. Figure 5.17 shows a typical schematic layout. Sometimes it is possible to dispense with a thermal storage tank if it is deemed that the thermal mass of the concrete floor is sufficient, once heated, to maintain a reasonable and consistent temperature throughout the year, day and night. An insulation layer with a minimum depth of 23cm (more than 9 inches) should be installed underneath the concrete floor, and thermal bridging minimised throughout. One example is a warehouse with integrated office for Neudorfer, in Rutzemos, Austria, built in 2005. The office section is built according to Passivhaus standards with an annual space heating consumption of 18kWh/m².

Table 5.2 The delivery temperature of the heat required to effect the same degree of subjective comfort, using different distribution systems. Note that the higher the temperature to be reached the more energy is required, disproportionally, and that the larger the surface area of the heat emitter, the lower its temperature can be.

Distribution system	Delivery temperature (°C)	Delivery temperature (°F)
Underfloor heating	30–45	86–113
Low-temperature radiators	45–55	113–131
Conventional radiators	60–90	140–194
Air	30–50	86–122

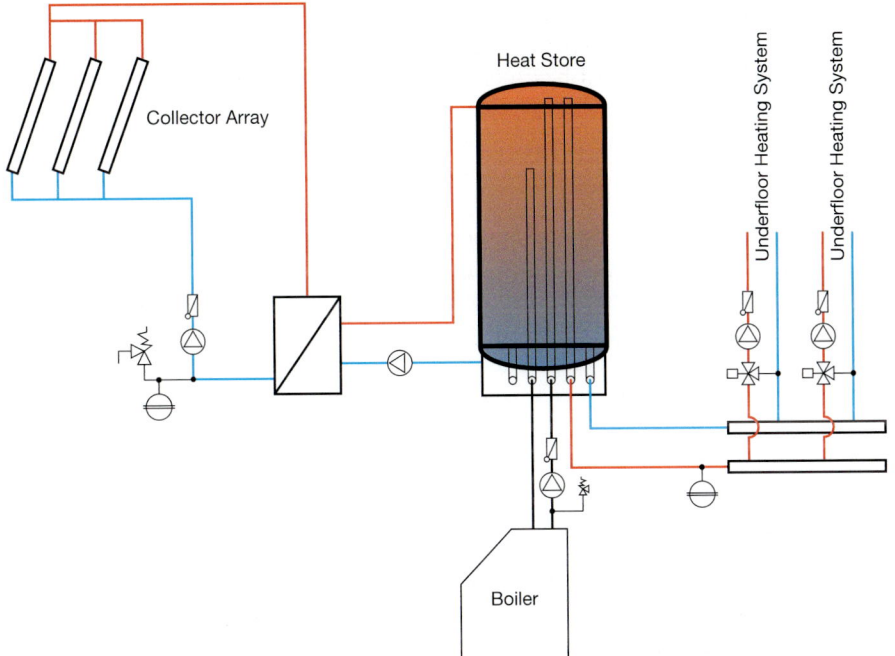

Figure 5.17
A schematic diagram for heating system using a combination of a boiler and a solar collector array to supply underfloor heating.

Source: IEA-SHC

Solar thermal collectors installed on the roof feed into an underfloor heating system, while façade-integrated PV panels supply electricity.

Heat stores

Whatever the heating source, there may be benefit from installing a heat store, namely insulated tanks that can store large quantities of heat in the form of hot water. They can be very useful additions to heating and cooling schemes,

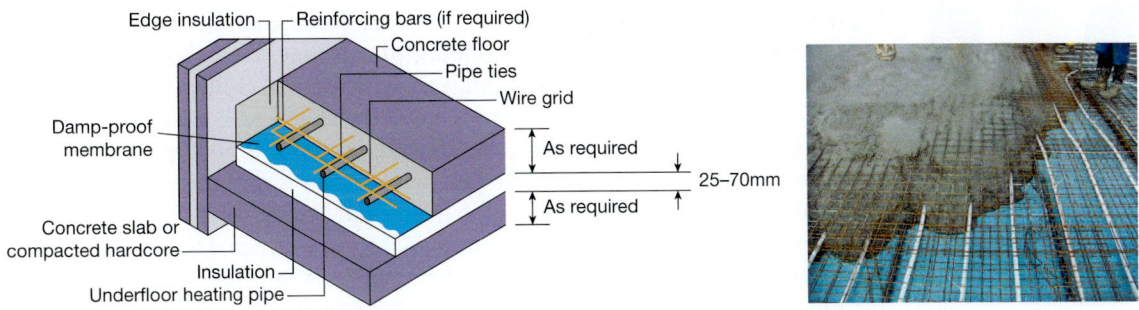

Figures 5.18a and b Installation of an industrial-scale system, as used in hangars and factories, based on the schematic diagram in Figure 5.17. An array of solar collectors on the roof feeds an optional, suitably sized storage tank supplemented by a boiler. This serves an underfloor heating system, where the pipes are laid in tandem with a reinforced concrete wire grid. The photograph shows an aircraft hangar in Austria.

Source: IEA-SHC

decoupling the direct link between the production of electricity and heat and its supply. Although often used in combination with CHP, heat stores may be combined with any form of heat-generation technology, such as heat pumps or biomass.

Thermal destratification

Thermal stratification is caused by hot air rising into the ceiling or roof space because it is lighter than the surrounding cooler air. It occurs in all buildings and can create dramatic differences between the ceiling and floor of a room. Where HVAC systems are fitted, this means extra work for them. A destratification fan will capture and reuse this heat. They are effective in non-air-conditioned spaces such as warehouses and factories, where internal environments can be cold during the winter months, even though heat generated by machinery, processes and people may be present in the roof space. The key to efficient destratification is delivering continuous, direct, non-turbulent airflow with maximised throw distance so that high- and low-level temperatures are fully mixed in a controlled fashion, using low-energy motors. Fractional amp draw axial turbine fans are recommended for this purpose. The leading brands begin operating on only 12 watts and can be angled up to 90° off vertical to direct the flow of air where it is required. Standard fans generally use more energy and cannot push air beyond 3 metres.

In the UK, major retailers such as Morrisons, Tesco's, Sainsbury's and John Lewis are using thermal destratification systems, saving significant amounts in operational heating and cooling energy. They can improve freezer and chiller aisle comfort levels with no impact on fridge or open freezer cabinet performance. Morrisons now specify thermal destratification in all stores. The United States Navy conducted research[2] which found a 40 per cent reduction in energy consumption in two facilities implementing thermal destratification fans. Short payback periods, of well under one year, have been reported.

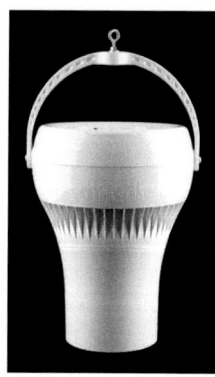

Figure 5.19 Modern thermal destratification fans direct hot air from ceilings to where it is needed with minimal electrical usage. New models may be wirelessly controlled, making installation easier.

Source: Airius

Solar cooling

Solar power is an excellent source of energy for cooling applications such as air conditioning, since the greater the amount of sunshine, the greater the demand for cooling. There is a wide output range of solar thermal chillers available, from 10kW to 5MW. They can therefore provide cold water as well as air conditioning. There are several technologies for solar cooling: absorption, adsorption, and solid and liquid desiccant. What they all have in common is that the external energy required to drive the process can come from solar heat.

Absorption chillers cover the entire 15kW to 5MW range. Here, a heat-driven concentration difference moves the refrigerant medium (usually water) from the evaporator to the condenser. The market availability of absorption chillers is mainly applied in combination with district heating or heating from cogeneration. Their coefficient of performance (COP) varies between 0.7 and 1.1.

Adsorption chillers, operating with a solid adsorbent, generally cater for the 50–400kW range. The input temperature of about 60–90°C (140–194°F) is provided by flat-plate or vacuum tube collectors with a coefficient of performance

Figure 5.20 A winery in Tunisia, which uses solar refrigeration. The solar collector at the front drives the absorption refrigeration machine. The wine in the fermenter tanks in the background is cooled by a cold accumulator.

Source: © ISE

of 0.5–0.7. They cost considerably more than absorption chillers, and need about 3–3.5m² of collector surface per kilowatt of cooling capacity.

Desiccant and evaporative cooling systems achieve cooling capacity in the lowest 20–350kW range. The operating temperature is only around 45–95°C (113–203°F); this means that the heat can be provided by simple flat-plate collectors and in some cases even air collectors, with a COP of 0.5–1.0 (the higher the COP, the better). The latest developments are triple effect absorption chillers with COPs of over 1.8. These systems supply cooling and dehumidification, and are widely available, with a high up-front cost but low operational costs.

In 2011, worldwide, about 750 solar cooling systems were installed, including installations with a low capacity (<20kW). More recently, a number of very large installations have been completed or are under construction, such as the system at the headquarters of the CGD bank in Lisbon, Portugal, which has a cooling capacity of 400kW and a collector field of 1,560m²; and a system installed at the United World College in Singapore, completed in 2011, with a cooling capacity of 1,470kW and a collector field of 3,900m². This installation is reportedly fully cost competitive. It was implemented with the help of an Energy Services Company (ESCo) model, under which the customer is not exposed to equipment or project costs; instead the ESCo sells the resultant cooling capacity to the customer. The flat-plate collectors are run at a temperature of around 100°C (212°F) to guarantee that the cooling unit has a nominal working temperature of 88°C (190°F). A transparent Teflon sheet between absorber and glazing reduces heat losses at high temperatures.

Some systems use a cold store, where cold water is stored to be used when solar power is not available to power the chiller. Solar cooling works well with a heat pump, as the heat may be returned to the ground or air. The heat pumps may then be used in reverse in the winter to heat the property.

Note

1 *Final Report: In-situ monitoring of efficiencies of condensing boilers and use of secondary heating*, Energy Saving Trust, June 2009
2 *Thermal Destratification Technology at West Bethesda, MD*, April 2010 – International Energy Agency.

6

Heating, ventilation and air-conditioning systems

Having considered the components of heating, ventilation and air-conditioning or cooling solutions separately, we are now in a position to consider the relative merits of integrated systems. Heating, ventilation and air-conditioning (HVAC) systems are integrated mechanical systems for maintaining an even temperature throughout the year in a building.

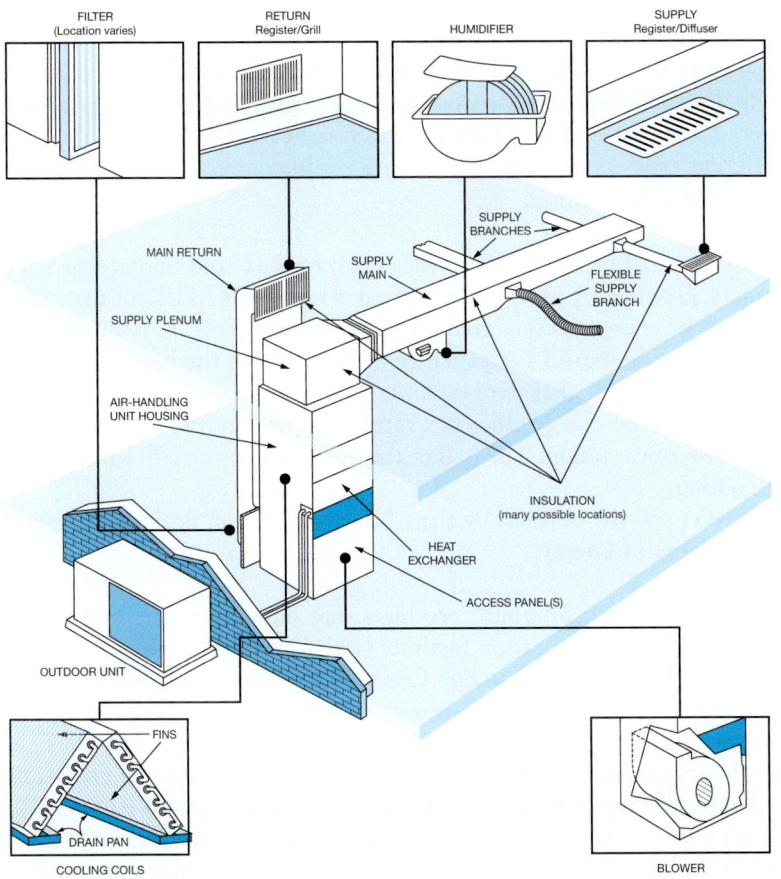

Figure 6.1 Components of a typical small heating, ventilation and air-conditioning system. The unit incorporates a heating source, which can be of any type.

Source: Author

Choosing new systems

If the existing HVAC system is more than 10 to 15 years old, a new one will be at least 25 to 50 per cent more efficient, and so it is worth considering replacement. Before choosing a system try to minimise the need for heating and cooling using all of the advice in this book thus far.

Many types of HVAC system are in operation, and there are a number of choices to be made, for instance, for cooling, whether water or a refrigerant is to be the coolant. They are typically installed and maintained under a service level agreement.

The key issues in choosing a system are as follows:

- sizing;
- zoning: yes or no? what areas?
- energy consumption and refrigerant direct emissions for air conditioners and chillers;
- energy consumption for terminal units and heat rejection units;
- heat reclamation;
- motor and fan efficiencies.

There are often potential trade-offs between heating and cooling modes.

System sizing is achieved by determining the degree day values throughout the year, and in accordance with specifications published by the US Society of Heating, Refrigerating and Air-Conditioning Engineers (ASHRAE) or the British Chartered Institution of Building Services Engineers (CIBSE).

Different types of heating and cooling equipment come with different types of energy efficiency ratings:

- AFUE (annual fuel utilisation efficiency): for gas- and oil-fired furnaces and boilers, given in terms of percentage. A 95 per cent AFUE means that 95 per cent of the energy is converted into heat while 5 per cent is wasted.
- HSPF (heating season performance factor): covers the heating function of a heat pump. The higher the HSPF the better the level of efficiency.
- SEER (seasonal energy efficiency ratio): for air conditioners and the cooling rating of a heat pump. The higher the SEER, the more efficient the system is at cooling.
- ENERGY STAR: covered by the US Environmental Protection Agency and Department of Energy.

It is recommended to purchase products that have been certified by the Air-Conditioning and Refrigeration Institute (www.ari.org) or which are on the UK Carbon Trust's Energy Technology Criteria List (http://bit.ly/10tNZGT).

In the USA, HVAC engineers are generally members of the American Society of Heating, Refrigerating, and Air-Conditioning Engineers (ASHRAE), which defines the standards. Under the ASHRAE 90.1 2010 standard, there is an allowance on the MEPS requirements for variable refrigerant flow (VRF) systems with a heat recovery function (which may imply supplementary head losses when operating in cooling or heating mode). In the UK, HVAC engineers are generally members of the Chartered Institution of Building Services Engineers (CIBSE), which publishes several guides to HVAC design relevant to the UK, the Republic of Ireland, Australia, New Zealand and Hong Kong.

Commercial buildings, operating from 8 a.m. to 5 p.m., five days a week, will have different needs from a manufacturing facility, an institution or a medical facility. This impacts particularly upon the choice of controls: programmable or fully automated. A fully automated system is almost always proprietary in nature, and can only be serviced by the vendor, unless the energy manager has received training, and it is part of their energy management system. With proper zoning and programmable thermostats, most projects will perform well and may be serviced by any maintenance arrangement. Energy management systems are deployed when the air-conditioning system is too complex to control with timers or thermostats. The systems, linked to the entire BEMS, allow for the use of different cooling temperatures for different zones, optimum equipment start and stop times, etc. Energy management systems can save 30 to 40 per cent on annual investment.

Zoning

Zoning is used if different parts of the building require heating or cooling at different times, or require different amounts of passive heating or cooling. The building is divided into different zones, each with its own system of controls and heating or cooling pattern, such as lower temperatures in unoccupied areas or different heating times. This applies especially to multistorey buildings, multiple buildings served by the same boiler house, shared buildings and multipurpose buildings. Before doing this, it should be checked that the installed equipment can cope with zoning. It is appropriate for large or complex buildings, but staff must be available to manage it.

Motorised valves are used to control water flow from the boiler to the heating and hot water circuits. Two-port valves may be used to provide zone control. Three-port valves are used with controls such as weather compensators. Thermostatic radiator valves (TRVs) can also assist with zoning. For larger buildings and sites, separate pumps and pipework would be required. A chief benefit is increased staff comfort and productivity.

Motors

Motors (pumps, fans, etc.) are likely to be the biggest energy consumers in the whole system. Choose variable-speed motors that meet the US National Electrical Manufacturers Association's Premium specification or those of the Institute of Electrical and Electronic Engineers (IEEE). Motor efficiency is of even greater importance in industrial facilities than in commercial buildings. Free software called MotorMaster+ to enable the comparison of motor systems for energy efficiency is available from the US Energy Efficiency and Renewable Energy office at http://1.usa.gov/HiaJ07. From the same address one can download calculators for variable-speed drives for fans and pumps.

Chillers/air conditioning

The system should use economisers, which operate at times when the outside temperature is lower than the inside temperature. They take fresh air from the outside for cooling rather than using refrigeration equipment to cool recirculated air.

Air-conditioning systems are best evaluated on the basis of seasonal energy efficiency ratio (SEER) metrics, as opposed to the energy efficiency ratio (EER). EER, a blunt instrument, quantifies the efficiency of the unit at maximum cooling output under design conditions. SEER, more precisely, is a weighted average of the energy efficiency ratio of the unit at different outdoor temperatures and cooling load ratios. A chiller should therefore be chosen on the basis of part-load performance, rather than when operating at their full-load rating. In practice, most chillers operate at full load for less than 5 per cent of the time. Since efficiency decreases significantly as the load on the chiller declines, full-load ratings will be of little use. Chilled water/boiler systems will require a higher level of maintenance for the cooling tower, boiler/furnace, heat exchanger and so on than a direct expansion system.

Variable refrigerant flow

A variable refrigerant flow (VRF) system is designed to minimise efficiency losses found in conventional HVAC systems. It may be thought of as a hybrid system incorporating a heat pump. It is based on the principle that a heat pump system may be designed to cool the building too if allowed to run in reverse. VRF uses a refrigerant as the cooling/heating medium instead of water, and allows one outdoor condensing unit (the coil in the ground or in the air) to be connected to multiple indoor fan-coil units, affecting different spaces, which are individually controllable by the user. By operating at varying speeds they work only at the rate needed, thereby saving energy. They may additionally provide heat recovery, whereby heat rejected by units in one area that needs to be kept cool can provide heat to an area that needs to be heated.

With water-based systems and with VRF systems, it is possible simultaneously to heat and cool different parts of the same building. In both cases, free heat, normally released to the atmosphere, can be recovered for simultaneous heating application, in some cases at a temperature compatible with sanitary hot water production. Adding a heat reclaim condenser in parallel to the chiller or air conditioner refrigerant condenser will recover heat for space heating, sanitary hot water or other heating uses.

All aspects of HVAC systems are being evaluated under the European Union's Eco-Design Directive to improve the environmental performance. Recommendations are not yet finalised at the time of going to print, but it is recommend that air conditioners and chillers have a large potential for improvement of all components if the best available technologies (BAT) are used.

Refrigerant choice

The environmental impact of HVAC systems is not only about energy efficiency but the type of refrigerant used, which could have a global warming impact when leaked into the environment as it invariably is. For air conditioners, there is no perfect alternative refrigerant for split and VRF air conditioners. For chillers, refrigerant choice is larger, because there is no refrigerant fluid circulated inside the occupied rooms, only chilled water. In both cases, the industry is now envisaging lower GWP (global warming potential) refrigerant fluids, in anticipation of a worldwide HFC (hydrofluorocarbons) ban. Replacement solutions envisaged include refrigerants which are pure HFO (Hydrofluoroolefins) refrigerants (flammable, but already used in car air conditioners), or blends of HFO and HFC refrigerants (currently the preferred option).

Much of the above applies to wet heating systems. Warm-air systems are relatively simpler. Controls mainly include a programmer, room thermostats, temperature sensors and controls on the heater, monitoring its firing and fan speed. Air systems should be linked to the air-conditioning system with heat recovery from the exhaust air.

If replacing components of the system, such as the chiller alone, this can accidentally lead to increases in energy use, especially when run at part load. The best approach for the energy manager is to look at the efficiency of the entire system as it runs in practice, rather than its components in isolation. This is to ensure that the system is efficient when run, as it often is, at part load as well as full load. The simplest way to approach this is to use a building energy performance simulation package. This would make it possible to model performance at all times of the year and the different loads. Then the best combination of controls, fans and chillers may be chosen. This type of software is offered by some HVAC equipment manufacturers.

As part of this system, controls must be used to properly sequence the chillers and match the temperature of the chilled and the condenser water to the outdoor conditions in real time. Tubes and cooling towers must be inspected regularly and cleaned, descaled, etc. For those systems which use cooling towers, the most efficient use to make of them is to run the condenser water over as many as possible at the lowest possible fan speed. This means that fans driven by variable-speed drives must be used. The ENERGY STAR advice is, for such systems, to 'open all the condenser-water isolation valves at the cooling towers and leave them open. To avoid additional pumping power costs, run only enough condenser-water pumps to maintain adequate flow through the chillers.'

Case study: Chiller replacement, San Diego Crime Laboratory, USA

The chiller plant at the San Diego Crime Laboratory operated with an energy cost of 1.48kW/ton, using two 130-ton air-cooled reciprocating chillers. The outdated equipment needed replacing. The chosen system was an all-variable-speed water-cooled chiller plant which included variable-speed cooling tower fans and pumps on both the chilled- and condenser-water side, as well as a magnetic-bearing compressor that could operate at variable speeds. This ended up performing with an average efficiency of 0.538kW/ton, with a payback of five years.

Optimising and maintaining HVAC systems

Faced with optimising and maintaining an existing system, a methodical check of all aspects of the system should be followed, prioritising those actions that will have the most cost-effective results.

Heat pumps

Installing a heat pump to help with heating and cooling is highly recommended, as it can boost efficiency by up to three times (see Chapter 5).

Thermostats

Thermostats should not be located near doors and windows, and must be set at the right temperature, usually 18°C (65°F) to 20°C (68°F). They are often set too high. It may be necessary to talk to staff to rectify any misconceptions that a higher set temperature means that, for instance, rooms will warm up more quickly. In offices and similar environments where heating is controlled manually, thermostatic radiator valves (TRVs) should be fitted to individual radiators; these provide the greatest degree of manual control. They should not be located near thermostats, as their operation can interfere with each other. They may be tamper-proof and locked at a fixed setting to prevent staff from interfering with them.

Control systems

The control systems can become desynchronised from the building's operational requirements. Identifying and rectifying the settings will immediately produce energy savings. Frost-protection setpoint temperatures may be set too high, which will cause them to run needlessly when the building is not occupied. Timing schedules may be set inappropriately for actual occupancy patterns, and even set for the wrong time, date and season. In air-conditioned buildings it is often found that heating and cooling systems are set to run simultaneously, in conflict with each other, and perhaps independently of the weather. They should be set so that there is at least 5°C (9°F) difference between when one cuts out and the other kicks in. It is often worth investigating the set operating times of ventilation systems to see if they correspond to the requirements of the occupancy pattern.

Any control system relies on the sensors sending accurate and pertinent information. They should therefore be situated in the most relevant place, and be calibrated correctly. Sometimes this involves conducting independent measurements to verify accuracy. A portable data logger would be necessary. The need for this could be identified by noticing that the heating system is systematically keeping a room at the incorrect temperature.

Pumps and fans

Pumps and fans may be incorrectly specified. The energy consumption from HVAC applications such as fans and pumps rises with the cube of the flow. Often

Figure 6.2 Variable-speed-drive-controlled HVAC systems come in a variety of sizes.

Source: ABB

they run at a fixed speed, which is frequently not matched to the requirement. Fitting variable-speed drives to these motors allows their speed to track the required demand, and is highly recommended to achieve considerable financial and energy savings. Energy savings of between 5 and 48 per cent are possible. (For more on variable-speed drives, see Chapter 8). Sophisticated drives come with a range of energy consumption monitoring and control systems. Drives can even be purchased as a series of modules with optional plug-ins such as a fieldbus for integration with the BEMS (building energy management system), a local control panel, mains disconnect, and so on.

Sequencing

Often, two or more boilers are servicing the same space, together with chillers or cooling towers. It is worth checking that, at any given time, especially in times of non-peak demand, the minimum number of boilers and firing times are being used for the purpose. When not in use, they must be isolated from the system, as otherwise they can still absorb heat and waste energy. Proper sequence control avoids short cycling, where the boiler keeps firing to top up the system. Boilers should be inhibited from firing up when not required. Rotation of the boiler firing order evens out wear, prolonging the life of the system. If a condensing boiler or a combined heat and power (CHP) unit is present, it should always take priority. These should be sized to provide the baseload.

Sometimes, where the HVAC system is incorporated within the BEMS, there is a complex series of different control loops, which affect both heating for water and space. This can result in uneconomic operation of boilers. An audit of the BEMS will usually show up this and any other existing anomalies.

Figure 6.3 Well-insulated piping in an HVAC system.

Source: Steve Karg

Check for deterioration

All heating pipes and ducts should be properly insulated with no breaks, i.e. well sealed with permanent, good-quality seals. Systems should be regularly inspected for defective dampers, valves and actuators as part of the maintenance schedule, and they should be replaced at the earliest opportunity.

Case study: a district heat metering and billing system, UK

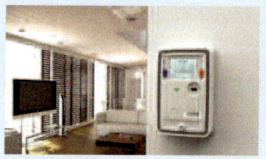

Figure 6.4 A district heat meter in situ.

Source: KNH

A metering and billing system was created for 1,036 homes spread across 28 housing schemes that are connected to the community heating supply of a local administration area in West Yorkshire, England. Kirklees Neighbourhood Housing (KNH, a not-for-profit company that manages Kirklees Council's council housing) commissioned the £1.7 million contract, which includes a five-year maintenance and data administration programme, from energy metering and billing specialist ENER-G Switch2 (ES2). This cost will be recovered over a long contract period as part of a fair service charge to residents.

In the first six months, residents have reduced overall energy consumption by 15 per cent; when winter consumption is factored in and residents get used to the system, KNH predicts that the annual average consumption will reduce by between 30 and 50 per cent. The system uses data collection and energy management technology for both pay-as-you-go and credit-billing consumers. It incorporates an in-home display that allows customers to see, in graphical form, how much energy they are consuming, when, the cost, and how much credit they have. Additional meters may be connected for potable water and electricity.

The pay-as-you-go element works much like topping up a mobile phone, using smart secure wireless GPRS technology to replace the traditional token-based prepayment system. There is a text alert facility to advise customers of low credit or to communicate service and energy-saving messages. The system avoids any requirement for an expensive building management system to be installed, or Mbus or other hardwired system.

Analysts continually monitor consumer data and can identify exceptional usage patterns or instances where residents frequently opt to disconnect their supply. This serves as an early-warning signal that can assist operators in targeting support towards residents who may not be using the system effectively. In one case, ES2 identified that an elderly resident's heating consumption had doubled over a 48-hour period. Staff established that he was immobile and were able to arrange immediate support.

'The key to reducing energy consumption is being able to see what you are using,' said Barry Goodwin, KNH Project Manager. 'With the economy of scale of supplying heat via a district heat network, we already provide lower-cost heat, but now that residents can track how much energy they are using and are taking action to reduce consumption, annual heating bills, including winter consumption, could average £5 per week. Heating bills for a similar sized property would cost approximately 60 per cent more if residents were purchasing gas or electricity from one of the big suppliers.'

ES2 designed and developed the system, with specialist software support provided by ADI, and hardware design provided by Blueprint.

7

Energy reduction technologies

This chapter takes a quick look at power factor correction, voltage optimisation and fuel conditioning. Chapters 8, 9 and 10 take a closer look at savings to be made in motors and drives, refrigeration, compressed air systems, process controls, supply chain optimisation, resource efficiency and disruptive developments. The information here is only intended as guidance. All devices and processes are specific, and all data sheets, manuals and operating instructions should be followed meticulously, with special attention paid to health and safety.

Power factor correction (PFC)

Part of the job of the energy manager is to get the most from the electricity that is supplied. The efficiency of an alternating current electrical system can vary widely according to the power factor. To optimise this it needs to be checked, together with all the loads upon the system. The definition of the power factor is the ratio of useful power to the total power drawn from the alternating current electricity supply (whether it is from the grid, or generated locally).

Ideally, the power factor ratio should be 1.0, i.e. all of the power drawn is used usefully. The power factor is 1 when the voltage and current are in phase. It is 0 when they are out of phase by 90°. Power factors are usually described as 'leading' or 'lagging', depending on whether the phase angle of current is positive or negative with respect to the voltage. In practice, achieving a power factor of 1 does not happen very often or very easily. A power factor of 0.85 or below is considered poor. A factor of between 0.95 and 0.98 is acceptable. The power factor is measured using specialised metering. An electrical engineer is usually consulted.

A low power factor will not only increase bills, but reduce the life expectancy of electrical equipment which, to function properly, requires a higher quality, stable power supply. Improperly installed power factor correction can introduce harmonics which can cause overheating of equipment and electromagnetic interference, which might be detected by communication systems, telephones, radios televisions and computers. It may even lead to equipment failure. Compensating for this can also help to avoid voltage drops over long cables and improving efficiency in supply transformers.

Equipment which will need correcting for are inductive loads, which include motors, induction heaters, fluorescent lights and welding sets. The most accurate way of identifying them is through a power factor survey. This is conducted by power engineering specialists and electricians. Sometimes utility companies and suppliers will also perform the service. Some equipment nameplates or manuals include power factor specifications, which may either be described as a figure or as an angle, or the cosine of the angle.

The power wasted in this context is called reactive power. Generators connected to the system are generally required to support reactive power, but some suppliers charge their customers a 'reactive power charge', or 'poor power factor penalty'. This is because they must compensate due to the excessive voltage drops that a poor power factor causes in the network. Thus, improving the power factor will reduce the energy bill. In Great Britain, the Grid Code requirements specify that the power factor must be between 0.85 lagging and 0.90 leading.

To balance the power factor, equipment such as shunt capacitors, shunt reactors, static VAR compensators and voltage control circuits are installed in circuits. The simplest form of 'fixed' PFC involves connecting capacitors in parallel to the loads. These can reduce the angle so that a lagging power factor gets closer to unity. However, it is possible to introduce too much capacitance to overcompensate, resulting in a 'leading' power factor.

Active PFC is an electronic system that controls the amount of power drawn by loads to optimise the power factor. It does so by controlling the input current to the load using power electronics. Sophisticated systems may be customised for complex sites. These are typically sites where many machines are being switched on and off at various times, causing the power factor to fluctuate. Any PFC equipment that has been installed needs to be regularly checked. This is especially the case if new loads have been introduced since the original equipment was installed.

Voltage management or optimisation

Voltage management, sometimes known as optimisation, stabilisation, regulation or reduction, reduces the voltage of electricity supplied to equipment in order to minimise consumption, and therefore bills. Minimisation must always be in line with the power requirements of the equipment.

The voltage of electricity supplied from the grid can vary above the nominal supply. This varies from country to country. In the USA it is 120V at 60Hz, but can vary from 114V to 126V. In the UK it is 230V at 50Hz, but can vary from 216V to 253V. In the USA, large appliances can use 240V. Large buildings frequently have 120/208V three-phase power, with large appliances being connected between two of the phases, giving a voltage of 208 volts.

Saving energy by voltage optimisation relies on the relationship between electrical power consumption, voltage and resistance. Power demand (kW) is expressed as a function of voltage: power equals voltage squared divided by resistance, or $P=V^2/R$. It follows that the higher the supply voltage, the higher the energy consumption. Because it varies as a function of the square, a 1 per cent increase in supply voltage will cause a 2 per cent increase in power demand.

The power consumption of voltage-dependent loads varies with the voltage being supplied to them. Another term for this is 'resistive load'. Not all equipment is voltage dependent; some is voltage independent. Examples include ICT equipment and electric kettles. A computer includes its own power supply unit, which gives a fixed output; the power supply to the computer is therefore independent of the supply voltage within a certain appliance-specific range. Similarly, an electric kettle will boil whatever the incoming voltage. If it is higher, it will just boil sooner. On the other hand, a traditional fluorescent lamp with an inductive ballast is voltage dependent. If more power is supplied it will consume it, resulting in higher costs (or a blown bulb!). A more modern fluorescent lamp with an electronic ballast is voltage independent. So are LED lights. Once this principle is understood, it may be possible to think of voltage-independent loads that may be substituted for dependent ones.

Fixed-speed electric motors are partly voltage dependent, and may be replaced with variable-speed drives which are voltage independent. Metal halide lamps with inductive ballasts and possibly process loads are voltage dependent. Any equipment controlled by thermostats, such as electrical heating or kettles, despite being resistive loads, are effectively voltage independent. An electrical specialist will be able to identify voltage-dependent loads under appropriate operating voltage ranges if there is any doubt.

Before instituting voltage optimisation, it is important to check all voltage-dependent loads to find the minimum voltage at which they will work. It is only worth instituting voltage optimisation if there is a significant number of these. The highest of all these minimum voltages is the one at which the mains input should be set. This is very important for safety reasons.

Example: A 10kW fixed motor is voltage dependent. If supplied at 230V, over eight hours it will consume 80kWh. However, if supplied at 242V it will consume 88kWh, an increase of 10 per cent. Over a year, this is a considerable amount of power.

There are three main strategies for voltage optimisation:

1   Installing site-wide or technology-specific voltage management equipment. There are two ways of doing this, depending on your loads: fixed ratio management provides a fixed voltage reduction at all times, for example, 8 per cent. Variable ratio management maintains a constant voltage regardless of the loads. This requires a thorough analysis and specialist help should be sought.
2   Adjusting incoming electrical supply infrastructure by reducing the voltage of transformers, or replacing transformers. Again, specialist help should be sought.
3   Replacing voltage-dependent loads with independent loads; see Table 7.1.

The steps to take for deciding whether to proceed are as follows:

Table 7.1 Voltage-dependent and -independent loads

Voltage dependent	Voltage independent
Fixed-speed motors	Variable-speed-drive-controlled motors
Lamps with inductive ballasts	Lamps with electronic ballasts
Uncontrolled heating	Loads controlled by thermostats (kettles, heating)
Process loads (possibly)	

1 Measure the incoming voltage and power.
2 Measure the voltage drops across the site.
3 Determine the proportion of energy consumption that is voltage dependent.
4 Identify any critical loads.
5 Calculate the potential energy savings.
6 Decide the power rating of voltage management equipment.

Voltage management devices can incur power losses as well, although they are low, so these should also figure in calculations. Once the potential energy savings have been calculated it will be possible to establish how much could be saved through voltage management and to compare this figure with the benefits of other energy efficiency projects.

Fuel conditioning

Fuel conditioning involves placing a series of high-powered magnets in a particular sequence on fuel feed lines to boilers that use oil, gas or LPG. They cause the fuel to burn at a higher temperature and more efficiently, although the exact physical cause is presently unknown. Evidence shows that fuel consumption can be reduced by at least 6 per cent, and sometimes by as much as 11 per cent. The savings made on fuel mean that payback on the cost of a unit is within one year, depending on fuel costs and usage. They work on all sizes of boilers.

When in place, the flame temperature increases by 12°C (54°F) for the same amount of fuel. This causes the heat exchange plate in the boiler to reach the target temperature sooner, meaning that less fuel is burned to reach the target temperature. The heat exchange rate, being hotter, causes the circulating water to be at a higher temperature so that when it returns it needs less heat to achieve the target. The bonding of oxygen to the hydrocarbon atom is also improved, so there is less waste hydrocarbon in emissions, causing less unburned fuel to be released into the atmosphere. The equipment is fitted in minutes by a specialist installer, with minimal disruption.

Figure 7.1 Fuel-conditioning magnets (yellow) around fuel feed pipes to a boiler.

Source: Magnatech Fuel Conditioning Ltd.

8

Motors, drives and compressed air

Motors, drives and compressed-air systems are examples of equipment that, over their lifetime, consume energy to a value far higher than the cost of the equipment. Therefore it really does pay not only to choose the most energy-efficient equipment in the first place for a particular job, but to have a proper maintenance programme for them in place.

Motors and drives

In the EU and North America, electric motor-driven systems account for approximately 70 per cent of total industrial electricity consumption.[1] This is probably typical of any industrialised country. Reducing their energy consumption would therefore save significant amounts of cash and carbon emissions. Motors have many purposes, whether in commerce or industry. They run pumps, machinery and compressors. They can be part of a larger system, such as an air-conditioning system or heating system. In such a case we can either talk about the efficiency of the whole system or of the motor itself.

Figure 8.1 A small variable-speed drive (VSD), or variable-frequency drive, which saves energy by replacing fixed-speed drives under the conditions outlined below; 50 to 70 per cent of industrial processes would benefit from VSDs.

Source: Wikimedia Commons (GNU Free Documentation License) C.J. Cowie

The cost of running an electric motor over its lifetime is hundreds if not thousands of times greater than its purchase price. Therefore it makes sense to choose the right motor or drive for every purpose. It also follows that the payback period for spending a little more to get the right motor is very short. The extra expenditure will pay for itself very quickly.

Since these motors can often be hidden away inside other equipment, they can easily be overlooked and left running when they are not needed. Part of the energy manager's survey or energy audit would be to identify such situations and develop strategies to remind people to switch them off, or to enable them to be switched off automatically when they are not required. Load-sensing devices are available for different applications to fulfil this purpose.

Sometimes one finds a pump in an air-conditioning system that runs continuously at a fixed level, with the airflow controlled by baffles (a valve-like device used to restrain the flow of a fluid or gas). The baffles should be removed and the motor replaced with a variable-speed drive (see p113). To further optimise a system like this, pipework or ducts should be sized to minimise flow velocities and friction losses for the desired purpose, and then choosing the appropriately sized pump or fan.

In examining the efficiency of an entire production system, it may be possible to redesign a manufacturing process to minimise the use of the motor, or to maximise its effectiveness when it is running. For pumps in liquid, air or gaseous circulation systems, the use of sensors that switch off a motor when a certain level is reached could substitute for valves that control liquid or airflow but keep the motor running at all times. See the section on process controls in Chapter 10.

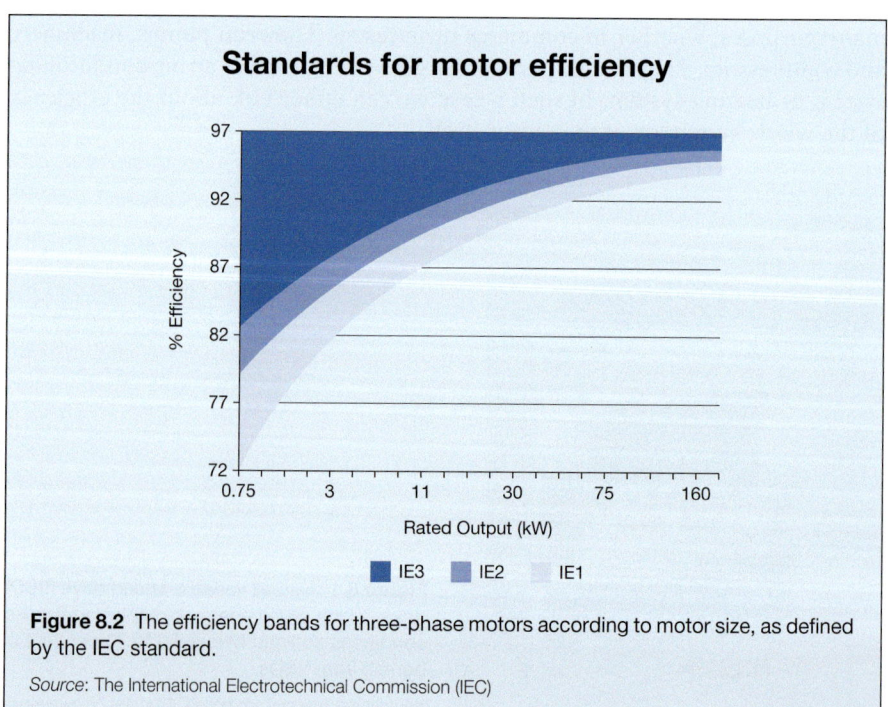

Figure 8.2 The efficiency bands for three-phase motors according to motor size, as defined by the IEC standard.

Source: The International Electrotechnical Commission (IEC)

Under globally agreed standards, motors are classified according to their efficiency level as follows:

Label	Efficiency level	Notes
IE1	Standard efficiency	
IE2	High efficiency	Same as the US Energy Policy Act (EPAct) set minimum efficiency levels for 60Hz; considerably higher than EFF2 from CEMEP (the Committee of Manufacturers of Electrical Machines and Power Electronics) for 50Hz.
IE3	Premium efficiency	Higher than EFF1 from CEMEP for 50Hz; in most cases identical to NEMA Premium in the US for 60Hz.
IE4	Super-premium efficiency	Higher than any other standard; not yet implemented.

The International Electrotechnical Commission (IEC) motor efficiency classification international standard IEC60034–30 gives a basic indication of efficiency for single-speed, three-phase, cage-induction motors with two, four or six poles.[2] In the above classification system there is a 3 to 4 per cent difference between each grade. An 11kW IE3 motor in continuous use will consume almost £250 less energy per year than an IE1 model. Presently, under European Commission rules, motors must meet the IE2 efficiency level. From 1 January 2015, those rated between 7.5 and 375kW must meet either the IE3 level or the IE2 level and be equipped with a variable-speed drive. From 1 January 2017, this will also apply to motors right down to 0.75kW.

In the UK the more efficient motors qualify for 100 per cent tax relief under the Enhanced Capital Allowance scheme. In the USA, for 60Hz operation, the IE2 and IE3 minimum full-load efficiency values are virtually identical to the North American National Electrical Manufacturers Association (NEMA) Energy Efficient and Premium Efficiency motor standards, respectively; this comes under the Energy Independence and Security Act of 2007 (EISA).
cf. Commission Regulation (EC) No 640/2009 implementing Directive 2005/32/EC

- NEMA standard: http://www.nema.org/Policy/Energy/Efficiency/Pages/NEMA-Premium-Motors.aspx.

Calculating the lifetime cost

The lifetime cost of an item such as a motor (although this formula will apply to many items) is calculated using the following formula:

Lifetime cost = capital cost + (n × annual running cost)

Where the annual running cost = (kW/eff) × L × hrs × elec cost
and:

n = years over which the payback is assessed
kW = rated kW of the motor
L = typical loading (how hard the motor is working in relation to its rated kW:
if unknown, use 0.75 as a default)
hrs = annual operating hours
elec cost = unit cost of electricity per kWh
eff = efficiency of the motor (%)

Example

A 15kW IE3 motor (which is 92% efficient) running for 4,500 hours a year at 71%
load consumes:

15 × 0.92% × 0.71 × 4,500 = 44,091kWh/yr @ 0.134p = £5,908.19 per year

If it lasts for 12 years multiply by 12 to give a total of £70,898.32 (not accounting for future increases in the cost of energy – which could be factored in at
3% per year).

If the motor cost £700, realistic at 2013 prices, it is now easy to see that its
capital cost is 1% of its lifetime cost.

The above motor is IE3 rated. If it was IE1, which has 87.6 per cent efficiency, and
had been rewound once, bringing it down to 85.6 per cent efficiency, running the
same figures through the above equation reveals that it would cost £6,308.45
($9,617) in one year, an increase of around £400 ($610). Given that a new motor
would cost no more than twice as much, it would pay for itself in under two years.

It is also now easy to understand that leaving the motor running when it is
not needed is extremely wasteful. A 15kW motor that is left running over every
weekend in a year would waste over £2,508 ($3,825) (52 weeks × 24 hours ×
0.134 pence × 15kW for the whole year (at the above rate, which represents 2012
UK prices).

Motors should be subject to regular maintenance checks, which include checking
the alignment of pulleys, belt condition and tension, lubrication and mountings,
loose terminals, whether the supply voltage is within the specified allowed devia-
tion from the motor's nominal rated voltage (+/–5% or so usually) and the line
voltages are balanced to within 1 per cent of each other. For motor controls and
force-cooled motors, the air filters in cooling or ventilation systems should be
checked and cleaned, or replaced according to the supplier's recommended schedule.

It may be beneficial to change or adjust a pulley or belt that communicates
the drive directly to the load, as, for instance, on a conveyor, compressor, machine

tool or other heavy industrial equipment. It is possible to achieve operating efficiencies of up to 98 per cent by installing them correctly. Flat belts or ribbed belts can typically yield savings of up to 3 per cent with a payback period of only a few years compared to V-shaped belts.

The most common type of motor is an AC induction motor. These usually run at a fixed speed. The most energy is consumed when a motor starts up. Motors also generate significant heat and have cooling fins to dissipate this wasted energy. Running them at a higher temperature is inefficient and renders them more liable to failure. It is therefore important to keep the fins clean and ventilated, and to replace them if faulty. Connecting a motor to a variable-speed drive limits the starting current and makes for a smoother start. It also prolongs the motor's life.

Variable-speed drives

A motor's optimum speed exactly matches that of the load. Variable- (or adjustable-) speed drives convert the incoming electrical supply, which has a fixed frequency and voltage, into variable versions of both, and allow the motor speed to be varied from zero to around 120 per cent of its maximum rated speed, so that it can match the load's power demand. Correctly designed VSD systems can reduce energy consumption by between 20 per cent and 70 per cent.

It is estimated that between 50 and 70 per cent of industrial processes would benefit from VSD. Yet, the current penetration rate of VSD as a proportion of installed motors is still low. Recent research by Siemens Financial Services has found that if the full potential of variable-speed drives (VSDs) was to be implemented throughout British industry, up to £2,512 million ($3,831 million) of energy cost savings could be made within the next five years.[3]

Using variable-speed drives on fans and pumps can produce significant energy and cost savings. What other devices can benefit? In general, there are three types of loads that may be driven by motors: constant power loads, constant torque loads and variable torque loads.

1   For constant power loads a variable-speed drive is not appropriate, as there will be no energy saving for any reduction in speed.
2   For constant torque loads, which include centrifuges, conveyors, extruders, grinders, mills, mixers, screw and reciprocating compressors, crushers and surface winders, the power consumed is in direct proportion to the useful work done; thus halving the speed will result in halving the power consumed. Other benefits include soft start-up of the equipment, reduced current on starting, reduced mechanical stress and high power factor.
3   For variable torque loads, since the power consumed varies with the cube of the motor speed, any speed reduction on the part of the motor will result in large energy savings. A 20 per cent reduction in motor speed can result in a 50 per cent power saving.

Applications which benefit most from variable-speed drives include the following:

- centrifugal fans and pumps that do not need to run at full capacity all of the time;

- HVAC ventilation fans for areas where occupancy varies;
- heating and chilled water circulation pumps;
- air compressors (where the average load on the compressor is less than 75 per cent and is frequently off);
- processes or systems driven by a centrifugal pump or fan where dampers and valves control the flow;
- extraction fans in dry areas;
- combustion air fans on large burners, where motorised dampers adjust the air-to-fuel ratio.

A variable-speed drive for a fan powered with a 5.5kW electric motor costing around £1,500 ($2,288) to install, excluding any controls, and operating for 80 hours per week, would therefore save £1,179 ($1,800) a year, based on electricity at 13.4p/kWh. The cost of purchasing the VSD would be recouped in just over a year. In this case it would be irrational not to replace the drive.

The following formula will calculate the energy saving of reducing the flow for a given fan or pump in this way:

$$P_2 = P_1 \times (RPM_2 / RPM_1)^3 = P_1 \times (Flow_2 / Flow_1)^3$$

Where:

P_1 = driven-equipment shaft energy requirement at original operating speed
P_2 = driven-equipment shaft energy requirement at reduced speed
RPM_1 = original speed of driven equipment, in revolutions per minute (RPM)
RPM_2 = reduced speed of driven equipment, in RPM
$Flow_1$ = original flow provided by centrifugal fan or pump
$Flow_2$ = final flow provided by centrifugal fan or pump.

Example

An adjustable-speed drive coupled to a motor delivers 20hp (14.71kW) to an exhaust fan when operated at its full rated speed. At a quarter of its rated operating speed, the fan delivers 25 per cent of its rated airflow, but requires only 1/64 or 1.5 per cent of full-load power. That is a saving of 98.5 per cent of energy cost. Even with the low drive efficiency of 47 per cent, with adjustable-speed operation the power required by the fan and the VFD is only 0.66hp (490W).

$$P_{25\%} = (20 \text{ hp} \times (1/4)^3 / (47/100)) = 0.66 \text{ hp}$$

Note: This example does not account for the efficiency at each load point for the fan-drive motor.

The use of VSDs can create harmonic distortion in the power supply. Where the load is small and the available power is large, this is not an issue. Where either a large number of low-current VSDs, or a few very large-load VSDs are used, this can damage equipment. Specialist advice should be sought. Before fitting a VSD on to a compressor, if it is made to operate at lower speeds, the oil pump and cooling fan will do the same. This could cause the compressor and motor to

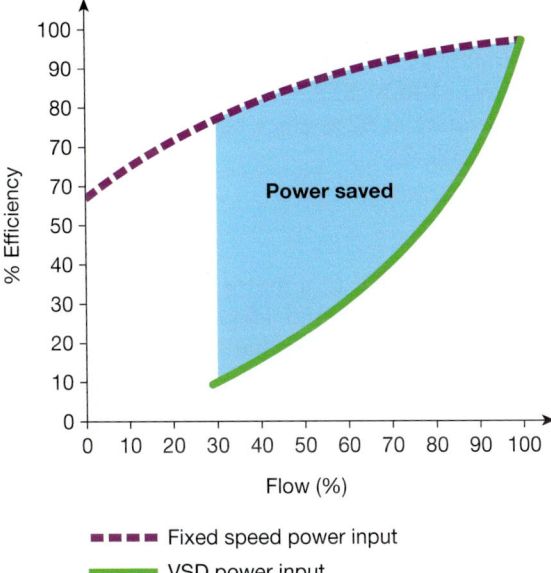

Figure 8.3 The area under the red line represents the power saved when changing a VSD pump or fan for a throttle-based one for a given flow. This is effective on applications where flow varies, such as, for pumps, circulating water in HVAC systems, boiler feed water and process pumps; and for fans, for ventilation systems, air-extract systems, industrial cooling, and combustion-air control systems for boilers.

Source: Author

overheat if it is not corrected. Modern VSDs contain processors that allow them to be connected to control systems and sensors.

Motor types and efficiency

Permanent magnet and reluctance motors can be even more efficient than induction motors. Hybrid permanent magnet motors combine induction and permanent magnet technology to replace conventional AC induction motors. Reluctance motors and drives are capable of very high-speed operation: up to 100,000rpm. In this situation they offer better performance than AC induction motors with variable-speed drives. DC drives can also be a viable alternative to variable-speed drives. They can provide high starting and accelerating torques. They all offer a higher power density, meaning they can be physically smaller for the same output, increased starting torque and a wider range of speeds.

The effect of loading

The 'loading' (or load factor) of a motor is the amount of work it does compared with its maximum rated power output. Operating a motor below 45 per cent of its rated output is usually inefficient. To determine whether such a motor should be replaced by a smaller one, you would need to compare the power each would actually draw powering such a load based on their efficiency. The manufacturer will supply a table or curve of efficiencies at different loadings. The power drawn is then calculated using the following equation:

Load factor = rated power/work done
kW drawn = rated power* load factor/efficiency

Example

A 100kW motor operating at 95% efficiency with an 85kW load, which gives it a loading of 85%, draws $(100 \times .85)/.95 = 89.95$kW.

A 75kW motor operating at 92% efficiency with a 62kW load would draw $(75 \times (62/75))/.92 = 67.39$kW.

Power optimisers

A power optimiser could be fitted to a motor which operates at below 50 per cent load. This can save between 5 and 15 per cent per year on energy costs. Power optimisers are solid-state devices that monitor the load and reduce the voltage accordingly. They are commonly used on equipment that is running continuously, such as fridges, freezers and refrigeration plant, air-conditioning equipment, as used in data servers, and manufacturing machines with cyclical loads with long periods of low-load operation or idling. They should not be used on motors with inverter variable-speed drives and VRV air-conditioning systems. Motors powered from three-phase electrical supplies generally also benefit from voltage optimisation in order to operate more efficiently.

Regenerative AC drives recover the braking energy of a load moving faster than the motor speed (an overhauling load) and return it to the power system, and are valuable in a situation where braking occurs.

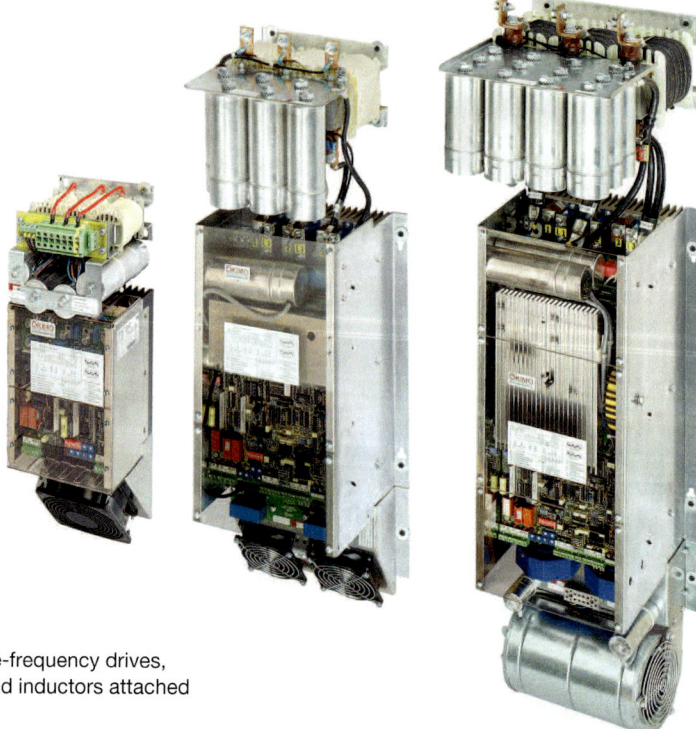

Figure 8.4 Line regenerative variable-frequency drives, showing capacitors (top cylinders) and inductors attached which filter the regenerated power.

Source: Wikimedia Commons, Dr Frank Oswald Hak

Maintenance

Motors can be repaired, which involves rewinding the coil. However, every time a motor is rewound it becomes up to 2 per cent less efficient. As a general rule, if a motor is below 5.5kW it should always be replaced rather than repaired. A motor should generally never be rewound more than twice. Usually, there is a justifiable saving by buying a new, higher-efficiency motor.

Installing monitoring (i.e. sub-metering), or using portable energy loggers on motor-dependent systems can help to identify when they become more inefficient and need attention. At the very least, they should be inspected every six months. Other ways of checking for problems on motors include: vibration analysis, where a probe is attached to the motor or gearbox and a frequency logger records vibrations at different frequencies; an analysis of oil in the gearbox; or thermographic surveys to identify overheating or electrical faults. A suggested strategy for monitoring and maintaining efficiency in motors and drives is as follows:

1　Create a spreadsheet of all of the motors and drives, and systems in which they reside. This would include their application, function, location, rated power, speed, efficiency class, power factor, loading, full-load current and repair history, plus hours of operation.
2　Prioritise systems for investigation.
3　Understand the objectives of each system and how it fits within the overall process.
4　Analyse how the motor contributes to this.
5　Model its optimum performance and whether the entire system or process could be redesigned to improve overall efficiency.
6　Could controls or power optimisers be a means to improve efficiency? Should it be replaced with a high-efficiency model?
7　List possible interventions and prioritise.
8　Determine how interventions can be monitored to measure any savings.
9　Develop the motor management policy with a maintenance programme.
10　Regularly check electrical energy consumption against baseline.

Sometimes, replacing a pump or fan with a higher-efficiency model does not result in observed energy savings. Usually this is due to the new model operating at a slightly higher shaft speed. The pump impeller or fan may need to be adjusted. If in any doubt, the supplier should always be consulted.

Compressed-air systems

In Europe and North America, 10 per cent of all electricity is used to make compressed air, so wide is its use in industry. Pressurised gases are employed for all kinds of processes and tools. As with motors, over a ten-year lifetime, the cost of the energy required to run a compressor is many more times its purchase price; in this case about four times greater than its initial capital cost. Maintenance also accounts for just 7 per cent of total cost but it can save energy in the same way as for a refrigeration system (which is covered in the next chapter).

A maintenance programme is essential, and should include the following:

- lubrication;
- replacement of air inlet and oil filters;
- cleaning oil and air coolers;
- checking operating temperatures and pressures;
- ensuring that filter pressure loss is kept to a minimum;
- replacing the desiccant at the recommended intervals where appropriate on desiccant dryers;
- checking drain traps are working correctly;
- checking for leaks or damage;
- checking operation of valves.

Sometimes compressed air does not need to be used at all, and another solution can be found for the task. For example, if the application is to pick up an item and transport it from one location to another, then a vacuum pump would use only one-third of the power than a modern compressed-air vacuum generator would to perform the same task. In another example, an electric angle grinder will use a quarter to one-third of the power of its compressed-air equivalent.

Sometimes the compressed-air-using tool needs to be upgraded or improved to save energy. Many blow guns are just open-ended pipes. The fitting of Venturi nozzles can achieve the same effect with less than half the energy. Operating a system at a pressure level higher than that required also wastes energy. It is important to ensure that the supply pressure matches that of the end usage. Investigating these two options has the potential for making the greatest energy savings.

Some education of staff can help to save energy. Sometimes compressed air is used for drying, cleaning or other purposes just because it is convenient. In fact it is an expensive way to clean surfaces and floors or dry products. Ensure that it is switched off when not required.

In large installations it may even be economic to institute an automatic sensing system that switches off the air supply when it is not being used. Companies making the biggest savings have trained staff to understand the cost of producing compressed air. All shop-floor workers must be made aware of the cost of air leaks and encouraged to report them as soon as they are noticed. Old industrial sites can have a leak rate up to 50 per cent. The following checks can be made along the line:

- That shut-off valves or condensate drain valves have not been left open or failed.
- That there are no leaks around hoses' jubilee clips, pipe joints or pressure regulators.
- That equipment is not left in operation when not needed. Leak detection equipment may be bought or hired to do this job automatically, but sometimes these can be detected when all other equipment is turned off, or seen with the application of soapy water.

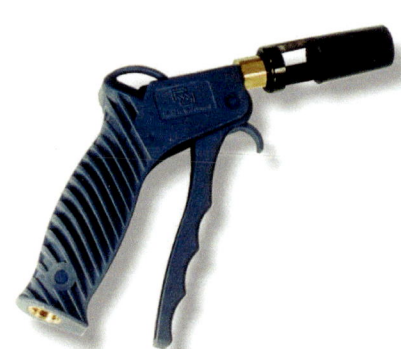

Figure 8.5 A lower-energy Venturi nozzle may be used for blow gun purposes instead of an open compressed-air tube. Just one example of where a simple, low-cost measure can make significant energy savings.

Source: Kiowa Ltd

Since a 3mm hole could cost over £1,000 per year in wasted energy, the cost of installing leak detectors can be recouped quickly. Monitoring of the system will help to detect any problems. A pressure gauge on the air receiver, data loggers and sub-meters on each compressor, airflow meters on the distribution system and temperature gauges in the cooling system will be of value here. It is useful to prepare a spreadsheet with records of compressed-air use plotted against production levels and energy consumed over time. Changes in any pattern established should be investigated.

Sometimes parts of the compressor system are not needed and contain areas of unused pipework which should be isolated when not in use. Air leakage will be minimised by reducing the pressure; it is worth checking whether processes can work just as well with a lower pressure setting.

Particular facilities will use other, more specialist equipment, besides motors, drives and compressed air. A small volume like this cannot possibly cover every particular piece of manufacturing equipment. The principles outlined in this chapter should be applied to this equipment as well to find further opportunities for saving energy.

Note

1 *Energy Efficiency Policy Opportunities for Electric Motor-Driven Systems*, International Energy Agency, 2011.
2 http://www.iec.ch/dyn/www/f?p=103:23:0::::FSP_ORG_ID,FSP_LANG_ID:1221,25.
3 Turn down the power, SFS Research Study, October 2012, http://sie.ag/158StBF

9
Refrigeration

Many types of business use refrigeration equipment. In most food production and food retailing businesses, refrigeration can account for 50 per cent of energy costs, bringing great potential for energy savings.

Purchasing new equipment

Refrigeration that is over ten years old should be replaced by more efficient models. As with motors and drives, the cost of running a fridge over its lifetime far exceeds its purchase price, so when buying new equipment it should be at the highest possible energy-saving specification. These use around 30 per cent less energy yet cost less than 10 per cent more than less efficient models; they often have tax rebates attached to them. In the UK, these are to be found on the Energy Technology List;[1] purchases are entitled to enhanced capital allowances. In the USA, they are to be found on the ENERGY STAR Qualified Products list[2] and Federal Energy Management Program[3] list.

When purchasing new equipment, models using propane should be strongly considered, as they give energy savings of between 10 and 15 per cent. Blends of pure, dry 'isopropane' (R-290a) (isobutane/propane mixtures) and isobutane

Figures 9.1a and b Supermarkets and other retailers are a major user of refrigeration, which frequently accounts for around half of their energy use. With no cover, the front-open-style freezer will lose a lot more cold than the top-access closed freezer.

Source: Author

(R-600a) have negligible ozone depletion potential and very low global warming potential compared to conventional refrigerants. They can replace R-12, R-22, R-134a, and other chlorofluorocarbon or hydrofluorocarbon refrigerants in conventional stationary refrigeration and air-conditioning systems. Leakage from these cabinets is low.

Optimisation and maintenance to reduce energy use

Typically there are quick wins to be found in refrigeration that can reduce energy use. A methodical approach will ensure that nothing is left out and will begin with investigating whether refrigeration is being used unnecessarily. This will cover an audit first of whether it is necessary to refrigerate products, their quantities, and for how long they should be stored. It will examine when refrigeration equipment is not being used to make sure it is switched off. Fridges must equally not be overstocked as this also wastes energy. Cabinets should be arranged so that the warm air output from some units is not flowing on to the condensers of other units, or even being recirculated around itself. Radiant heat reflectors fitted above open-top freezer cabinets can reduce heat gains from them by around a quarter. This is most applicable to retail cabinets.

Fridges should then be set at the highest permissible temperature. Heat ingress should be reduced by, for example, keeping cold room doors closed and well sealed; this can save 30 per cent of losses. Night covers should be fitted every night where appropriate. Strip curtains or airlocks should be used, and insulated rooms must be properly airtight. Faulty vapour seals should be repaired so that moisture does not get into the cold room. Any insulation in bad condition must be replaced, and there should be as much insulation as possible. Insulation should prevent condensation forming on the surface. It must be airtight and thermally sound. All joints in the insulation should be sealed. Air pressure tests or thermographic photography are good ways of checking for breaks. They can also show up whether pipe insulation is sound and continuous.

Air grilles must be kept clear. If the grilles at the front of an open-fronted cabinet in a shop are blocked by produce, this lets cold air into the shop. This forces both the refrigeration plant and the heating system in the shop to work harder. Expansion valves, where fitted, should be properly commissioned, and of the electronic type.

Any interior lighting should be T5 or LED, as these do not give off heat (see Chapter 4).

Following this attention to the 'quick wins', the next step is to implement a maintenance system. This is usually carried out by contractors. The annual maintenance cost is between 2 and 5 per cent of the capital cost of the equipment, but up to ten times the maintenance cost over the lifetime of the equipment will be saved through improved energy efficiency, reduced service costs and lost production costs.

The maintenance programme should include the following:

1 Checking the liquid line sight glass, if fitted, to make sure the level is full.
2 Cleaning condensers and evaporators and having them serviced regularly, which can cut energy consumption by up to 10 per cent.

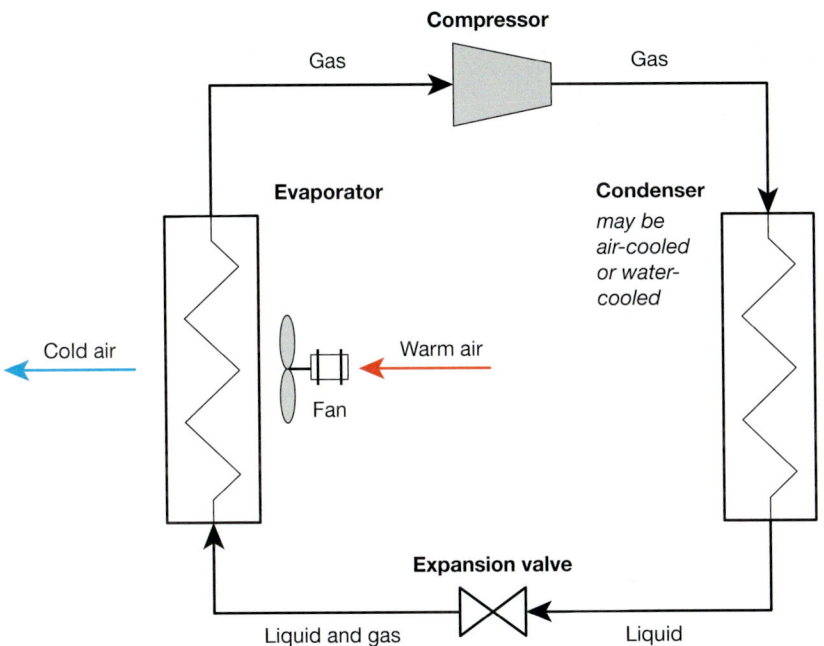

Figure 9.2 Layout of typical single-stage vapour compression refrigeration.

Source: Author

3 Recommissioning, which often has a dramatic effect.
4 Regular defrosting.
5 Finding and repairing leaks, which can reduce energy consumption by up to 15 per cent.
6 Inspection and repair of pipe insulation, which can save 10 to 25 per cent of energy.
7 Checking condenser fan and pump condition.
8 Checking condensate pipes are not iced up.
9 Checking for all the improvements that were made on the initial survey.
10 A monitoring regime that checks for any changes in pressure and temperature, which should be immediately attended to.

Compressors

Compressors lift the temperature of the refrigerant by increasing pressure so that heat can be expelled. They are the most energy-intensive components of the system. They should have a pressure control with a low-set condensing temperature in cooler weather. Between 2 and 4 per cent of compressor power can be saved for every 1°C reduction in condensing temperature, so it makes sense to reduce the temperature lift as much as possible by lowering the temperature at which heat is discharged, according to the daily or seasonal outside temperature. Typically, the condensing temperature is set at 40°C (104°F) all year round regardless of the weather. Reducing this to 20°C (68°F) in the cooler months and letting it track the outside temperature by fitting a thermostat will lead to 25 per cent energy savings.

Figure 9.3 It is bad practice to keep the condensing temperature constant throughout the year. Setting it at a much lower level but allowing it to track the external temperature when needed will save valuable energy.

Source: Author

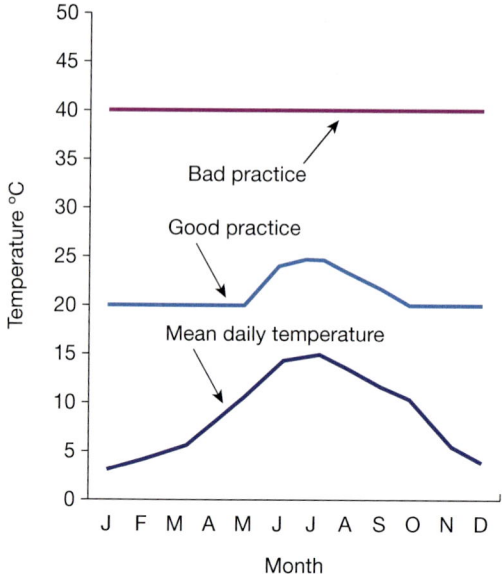

If it is not possible to reduce the condensing temperature, a solution called liquid pressure amplification is also available, which has a typical payback of three to five years. This service-level solution involves the installation of a pump in the outlet line of the condenser after the liquid receiver, to provide the stable pressure needed for the expansion valve. It permits the condensing pressure (and hence compressor delivery pressure) to 'float' with ambient temperatures. Although it requires energy to operate, this is far outweighed by that which is saved. This may also be applied to air-conditioning systems.

Reducing the head pressure set point from 15.1 bar to 12.0 bar can result in energy savings of more than 22 per cent in the summer. If the cooling load varies widely, speed controllers may be attached to the compressors. Any of these solutions can potentially save tens of thousands of pounds or dollars per year.

Evaporators

Evaporators should only defrost when necessary. Intelligent controls may be added which detect when a defrost is required to optimise energy use for this purpose. They must be the right size for the job. They should be regularly cleaned as part of the maintenance programme; if the evaporator is blocked or not controlled properly, cooling will be inadequate or excessive energy consumed.

Heat recovery

Up to 90 per cent of the electrical input to a refrigerator's compressor is wasted in the form of heat and may be recovered. This means that recovering the waste heat from air- and water-cooled compressors can save money and carbon emissions. Many types of compressor exist, such as the air-cooled oil-injected rotary screw, piston, oil-free screw, centrifugal and vane types. Waste heat may be recovered from all of these (Table 9.1).

Table 9.1 Savings potential from compressors. Savings based on replacement of gas heating with a boiler efficiency of 85 per cent and a gas cost of 4p/kWh.

Compressor capacity	Nominal motor rating	Annual heat available (2,000 hrs/year)	Savings potential
cfm (cubic feet per minute)	kW	kWh	
90	15	24,000	£1,130
125	22	35,200	£1,656
350	55	88,000	£4,093
700	110	176,000	£8,234
1,000	160	256,000	£12,046
1,250	200	320,000	£15,059
1,550	250	400,000	£18,824

Source: the Carbon Trust and author

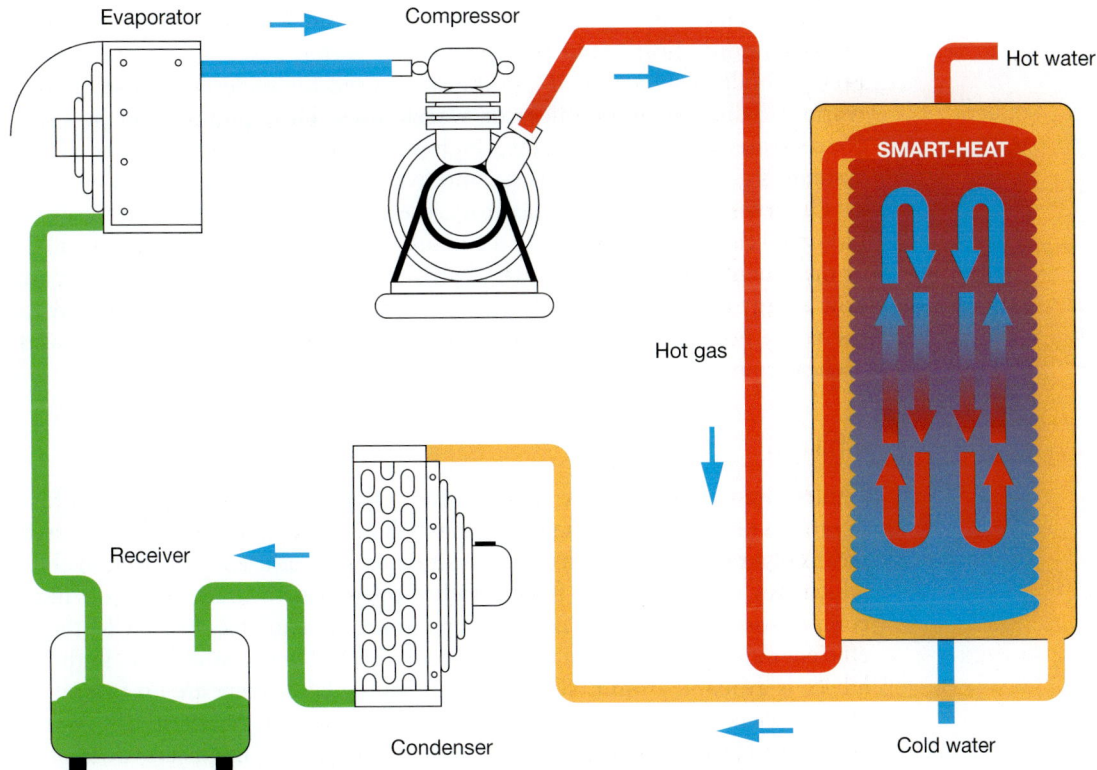

Figure 9.4 A schematic diagram of a heat recovery system from an evaporator (for example, on a refrigerator), which heats water.

Source: Tradewinds Engineering

It is worth considering fitting heat recovery systems to refrigeration. Most businesses also require heat, so the heat output from refrigerators can be transported to where it is needed, reducing heating bills.

The choice of recovery system depends on the compressor type and the target use for the recovered heat, for example, for general hot water use, or industrial applications. The quantity of heat that is available should also be known before designing and sizing the system. A survey should be conducted to investigate and determine an optimum end use for the heat. For example, it could be used to regenerate desiccant or sorption dryers, for process heating or preheating, for producing warm air to keep materials dry, or for space heating and hot water heating.

Temperatures as high as 80°C (176°F) may be obtained. The heat may be used directly in an air-based heating system, or transported to a hot water storage tank via a heat exchanger. If the heat recovery times do not match the demand times, heated water can be stored in insulated tanks. The compressor supplier should be consulted for safety reasons. A heating engineer should be contracted to do this work.

For example, a medium-sized supermarket could have a cooling load of around 230kW and about 310kW ejected into the atmosphere. Ten per cent of the heat ejected will be relatively high-grade heat available at 50 to 60°C (120 to 140°F). Even the low-grade heat, between 20 and 30°C (68 to 86°F) is useful. The high-grade heat could be used to heat water, and the low-grade heat to preheat boiler feed water. It could also be used for underfloor heating, or the warm air directed into offices or warehouses. Heat pumps may be used to concentrate low-grade heat into high-grade heat. The payback for all these systems is generally between three and five years.

Otherwise known as recuperators, these heat pumps have wider uses in an industrial context. The supply and exhaust air streams of an air-handling system, or the exhaust gases of any industrial process, can have a special purpose counter-flow energy recovery heat exchanger positioned within them in order to recover the waste heat. They are frequently used to recover heat from waste gases to preheat combustion air and fuel.

Chillers

Chillers are used to produce chilled water, glycol or another chilled fluid for use in process or air-conditioning equipment. Chiller efficiency is usually expressed by means of a ratio. Those with a higher ratio are more efficient. Absorption chillers, which can use waste heat from another operation to drive them, do not use compressors and should be used if such a source is available. In the UK, the most efficient models may be found on the Energy Technology List website. In the USA, advice is available from ENERGY STAR. In the USA, there are no federal minimum efficiency standards applicable to chillers. However, ASHRAE (the American Society of Heating, Refrigerating, and Air-Conditioning Engineers) does provide efficiency specifications in its 90.1 standard, 'Energy Standard for Buildings Except Low-Rise Residential Buildings', which is used in many local building codes.

Chillers used for air conditioning might use one of four possible types of compressor: reciprocating, scroll, screw and centrifugal. The first is the least efficient. The second and third are intended for applications requiring up to 300 tonnes of cooling capacity. The fourth provides for larger capacities.

Chillers are frequently selected and set to satisfy the maximum cooling load during the warmest weather. Most of the time they will not be running in these conditions. To optimise the functioning of chillers, the strategy is to determine the maximum load and ambient temperature required, and obtain from the supplier the expected performance details with a part load in cooler weather. A model which can adjust to the load and which makes the best use of heat exchangers should be selected. Adopting this approach, savings may be obtained in the region of £25,000 a year for an 800kW chiller. Since they may cost around twice this amount, there is a simple payback of two years. It is sometimes possible to integrate free cooling into the chiller, an option offered by several manufacturers. See Chapter 6 for more information.

Case study: Carrier's Charlotte LEED EB Project, USA

Carrier's Charlotte factory in North Carolina manufactures a high-efficiency chiller that, at its inception, was 40 per cent more efficient than the industry standard. The factory has earned a Leadership in Energy and Environmental Design Existing Building (LEED® EB) certificate by improving the energy efficiency of its operations. Among these were changes to lighting and temperature controls, and the installation of a heat recovery chiller and variable speed-drive air compressor. The initial investment of $528,000 was recovered through energy cost savings in under two years.

Notes

1 Energy Technology List: http://etl.decc.gov.uk/etl/find/.
2 ENERGY STAR Qualified Products list: http://www.energystar.gov/index.cfm?Fuse action=find_a_product.
3 Federal Energy Management Program: http://www1.eere.energy.gov/femp/technologies/ eep_purchasingspecs.html.

10
Process controls

Process control is at the heart of any mass-production process which operates on a continuous basis, from oil refining to paper manufacturing, processed foods, chemicals production, power plants and product assembly lines. Systems to control these have been evolving since the 1970s, but still adopt the same philosophy of approach. This is based around the process control hierarchy.

The process control hierarchy works from the bottom up. The first level is the installation of sensing, measurement and data collection. Then, closed-loop control systems look after planned operations automatically, requiring intervention only occasionally at normal times. However, because they are not designed to handle exceptional situations, for example when faults arise, they must be supervised either by human operators or higher-level systems. These, known as supervisory control and process diagnostics, are set and defined under the process conditions to help optimise energy use as well as production efficiency.

In a modern large plant, hundreds of individual feedback loops can comprise the whole system. Their individual jobs may be to control the temperature or pressure of an operation, the speed of a motor, the length of a process, the rate of flow of a liquid, and so on. They consist of a measurement device, a controller and a regulator. The controller, typically a piece of software on a PC device,

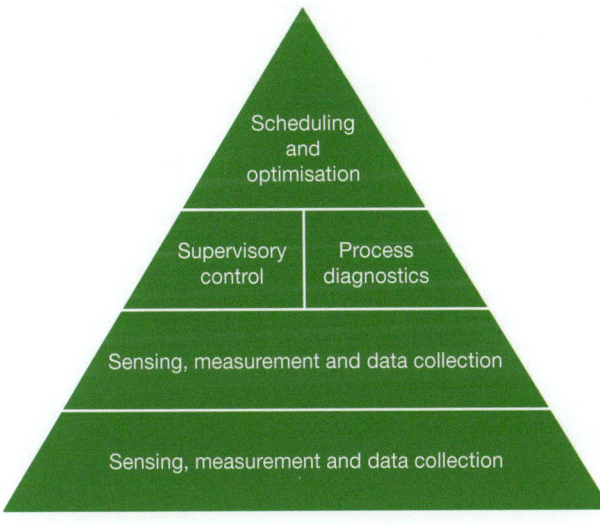

Figure 10.1 The process control hierarchy.

Source: Author

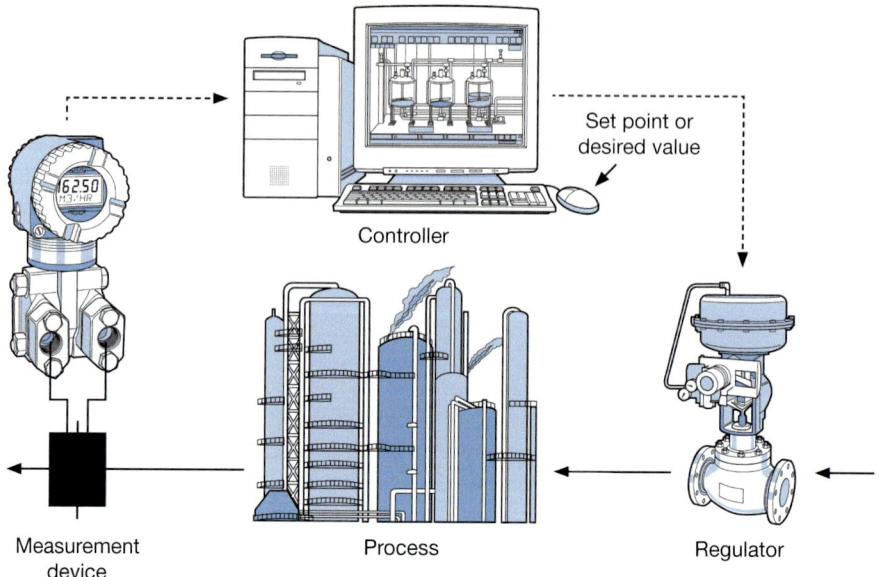

Figure 10.2 The main components of a process control loop.

Source: The Carbon Trust

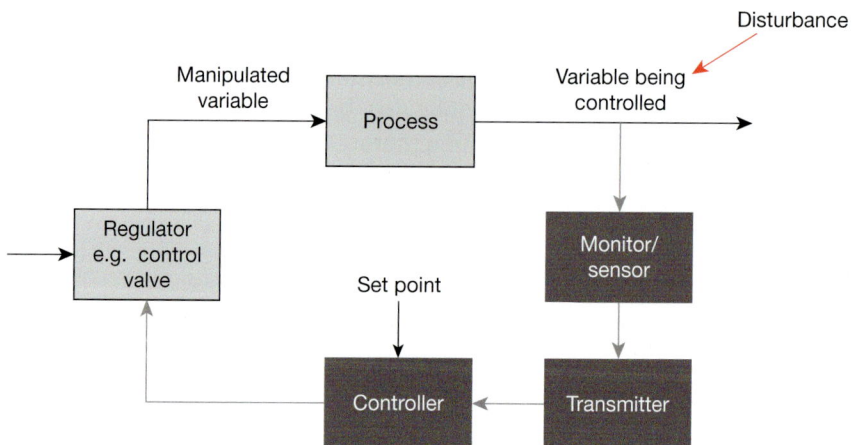

Figure 10.3 A block diagram of a single-feedback process control loop. A regulator of a process, such as a valve, is controlling a variable on the process. This may be the amount of an ingredient being added to a compound, for example. A sensor monitors whether the correct amount of the ingredient has been added and transmits the result to a controller. The set point of the controller is determined by the operator. If the sensor finds that the incorrect amount is being added it can adjust the regulator. Changing the set point would reprogramme the sensor as well as the regulator.

Source: Author

defines the set point of the operation and controls it with an algorithm. The monitoring equipment informs it of whether this point has been reached or whether it is necessary for the regulator to make an adjustment. If so, a signal is then sent to the regulator.

Case study: Attention to detail pays off

A processing unit had an impurity specification of 20ppm for a particular product, but the specification actually being delivered was generally below 4ppm. The product was therefore of excellent quality. However, energy consumption was excessive because the process control set points were not being adjusted to align the energy input to the feed flow. Since, when detected, the situation had remained unchanged for five years, this represented a significant energy cost. Altering the set point took the operator just three hours, yet it resulted in annual energy savings of £250,000 ($381,000).

Wireless controls

Lighting and heating controls which depend upon occupancy levels are a basic everyday example of such a system which we have already met. We have also discovered that, as the 'internet of things' develops, these devices can increasingly be controlled wirelessly, and from any point on the network. This is enabling faster implementation of a new generation of wireless process controls.

Figure 10.4 This photograph shows the difficulty in locating monitoring and control equipment. Wireless equipment is easier to install. At this site, researchers at the University of Texas at Austin have demonstrated closed-loop control of a distillation column using wireless technology, with performance virtually indistinguishable from that of wired transmitters.

Source: Emerson Controls

Figure 10.5 CalPortland, a US manufacturer and distributor of cement and concrete, installed a wireless device to monitor a rotating cement kiln at its plant in Colton, California. This rotates with the kiln and helps CalPortland meet nitrogen oxide emissions regulations and return savings in energy use and maintenance costs. The conditions made using a wired solution impossible.

Source: Emerson Controls

Figure 10.6 A variety of wireless process monitoring and control equipment which includes pressure, temperature, level, flow, vibration, discrete switches and pH devices, as well as wireless-enabling devices, valve position monitors, a secure Smart Wireless Gateway, and Device Manager software to manage predictive diagnostics.

Source: Emerson Controls

Modern control systems may therefore be integrated wirelessly and with less effort or disruption to the production line, a very important factor in persuading management to sanction energy efficiency measures in production lines. There is no need for the costly introduction of wires and cables. In large plants, cable routing can account for the highest expenditure when installing monitoring. Frequently, lost income from production line downtime is given as a reason why energy efficiency measures are not introduced. Wireless monitoring both means that there can be wireless network access that is useful for any other purpose throughout a plant, no matter how large, and means that the gathered data may be accessed anywhere in the world where there is internet access.

Back to control loops. A controller in a single-feedback loop can be used to adjust the set point of another controller. This is known as a 'cascade system'. It might be used, for example, to control inputs into a reactor chamber dependent upon the temperature and pressure inside. Nowadays, such systems are controlled by software with programmable logic controllers, whose software interface includes icons that, upon double-clicking, expand to allow variable set points to be changed easily, and modelled or optimised before being actuated.

Figure 10.7 Part of a wireless network that monitors the gas balance of a treatment plant for Tecpetrol, a natural gas producer in Argentina. Variables monitored include the dew point, primary separation, compression stages contracted, consumption and plant venting.

Source: Emerson Controls

Distributed control systems

Cascade systems can be modules. Putting several modules together produces what is known as a distributed control system. These may start up or shut down operations and apply advanced control techniques across multiple sites. Distributed control systems will include a high-speed network or control bus and a central control unit. They will contain measuring points which include sub-meters, inductive flow meters, and orifices for steam flow measurement wherever appropriate.

The energy manager will work with the controller of these systems in order to obtain information for their own auditing processes, and then to optimise the production line efficiency. Their data will be obtained via a supervisory control and data acquisition system, known as a SCADA. This includes the ability to store and distribute process data for analysis. The latest versions may also include functions that can automatically optimise process operations, identify faults and schedule production systems.

Integrated energy management with process management

Complete, integrated energy management is only possible if the relevant process information is acquired automatically and made available in a suitable form. It needs to be capable of permitting visualisation of the information, producing reports, archiving, and to integrate with the energy management software. This entire process is summarised in Figure 10.8.

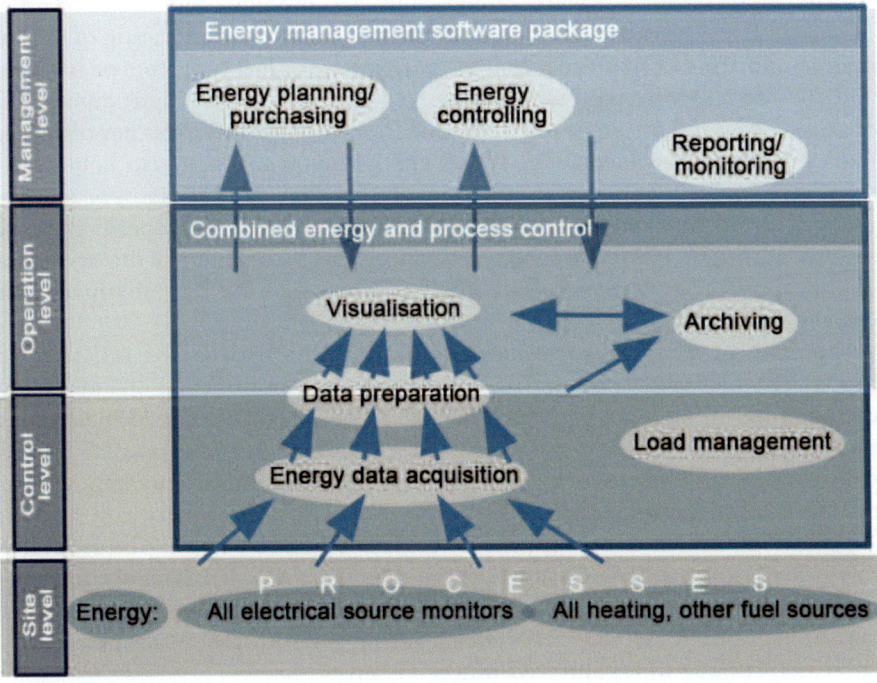

Figure 10.8 Integrated energy management with process management: a block diagram illustrating data flows in control loops. Modern solutions not only allow for optimisation processes to maximise efficiency in production and energy use, but also to track faults, produce audits, attach costs, and produce financial reports, enabling transparency of the entire operation up to senior management level.

Source: Author

Figure 10.9 Wireless monitoring in the process control room of a chemical plant.

Source: Emerson Controls

The most useful software applications are modules and functions for Microsoft Excel spreadsheets. These can provide automated allocation of energy consumption and costs to cost centres, particular production plants or batches, and considerably ease the burden of work. Being able to visually compare and analyse consumption across different time frames, using the archiving function, makes it possible to identify potential energy savings. It can also help avoid consumption peaks to manage total or maximum demand. Loads can be quickly dropped or bought in when required, for example, to work within specified limits, smooth peaks, shed unstable supply networks or loads, or continue the operation of critical processes while suspending others. These may work on anything from one-minute to half-hour reporting times.

In some proprietary software on the market a 'dashboard function' is available, which allows for traffic-light representation of deviations in energy consumption behaviour. These can be forwarded to predetermined individuals responsible for parts of the plant by e-mail. Information is also available via a web browser, and any users may be permitted access to this, for example, plant managers or clients and customers.

This is important for energy procurement, since load curves may be used later to positively influence energy supplier contracts. Very detailed data on consumption structures for energy accounting and planning are possible. Any variables may be included, for example: electricity, steam, gas, nitrogen, well water, external water, cooling water and feed water.

Maintenance of the process control system

Energy management that involves the 'tuning' of processes for optimum energy use can result in energy savings of between 5 and 15 per cent, depending on the quality of previous processes. In one survey, one-third of control loops were found to be set at inefficient levels. Even on the best sites no more than 70 per cent of their control loops performed well.

The positioning of a set point on each control loop needs regular monitoring. For example, if several different products are being produced on the same production line but they require different settings, the set point is not always returned after one batch has been finished. For example, a chemical product may require a certain value of purity, and this may be set higher than necessary. A process temperature may have been set too high because material going through it was damp and needed drying. Subsequent materials are not damp and do not require such a high temperature, but the process temperature was not reset.

A maintenance schedule should also be in place to check that every single regulator and monitor is functioning correctly, and automatic controls are not set to manual. Are monitors in the best position? Is there an acceptable time lag between monitoring intervals and a response to a change in conditions?

Checks will include an inspection of the cabling, if cabling is used. In certain situations cables can pick up interference from nearby electrical power cables or equipment. This introduces noise into their signal output. The result can be significant and produce inaccurate readings, or cause operators to switch to manual control. There are three solutions: introducing wireless communication; shielding the cables; and introducing a digital IEC 61158 Fieldbus, which means reduced cabling since many devices can share the same set of cables.

Tuning controllers and training operators are strategies with no capital cost and the payback period can be measured in hours. Switching to automatic control from manual almost always saves money and energy, and has a low capital cost with a payback period that is measured in days. Improving measurements and making changes to set points or algorithms and other simple changes have a medium level of capital cost that is repaid on average in under six months. More advanced solutions can take up to 18 months to recoup their costs, while a complete refit and upgrade of process control systems, which is advisable if they are over 15 years old, on average has a payback period of around two years.

Once changes have been made, the results should be monitored against performance targets to ensure that they are as expected.

11
Data centres

The modern world's reliance on computers and the internet depends upon the deployment of huge numbers of specialised computers called servers which store data and may respond to queries sent via a network or the internet. Collections of such servers are called data centres, and they consume a phenomenal and increasing amount of energy. Worldwide, this amounts to about 30GW annually, approximately the same as the output of 50 nuclear power plants. The large data centres, which serve the computing 'cloud', consist of thousands of rows of servers spread over hundreds of thousands of square metres. Many organisations have smaller ones, down to single roomfuls of servers, servicing their own needs. Data centres probably have the most potential for reducing energy use of any industrial or commercial sector.

Figure 11.1 A modern, neat data centre. The servers are positioned in racks, with cabling at the back entering the floor. Each rack can slide out. Empty racks should be covered. Note the raised floor, containing air circulation vents, to accommodate the cables.

Source: Author

Figure 11.2 A server rack in a data centre belonging to a website hosting company.

Source: Wikimedia Commons (Jfreyre)

Cloud computing is the term given to the storage of data on servers connected to the internet rather than on a local computer such as a PC, laptop or smart phone. According to research by energy writer Mark P. Mills at www.Energy-Facts.org, cloud computing consumes ten times more energy than storing it on a desktop PC. He has found that when a user accesses the cloud frequently, or performs high-intensity tasks, such as downloading music, video and playing games, over ten times more energy is used than if it was retrieved from a PC or laptop. Mills quantifies this by saying that downloading one gigabyte of data every day from the cloud (say, 114 minutes of uncompressed CD-quality audio, or one hour of standard-definition television) requires the energy of 3lb (1.36kg) of coal per year.

At the extreme, in Silicon Valley, California, many data centres are on the state government's Toxic Air Contaminants Inventory,[1] meaning they are top polluters owing to the banks of diesel generators used for backup. According to a survey[2] conducted by consultants McKinsey, many run their facilities at maximum capacity 24 hours a day, seven days a week, whatever the demand is on their use. This means that they waste 90 per cent or more of the electricity they take from the grid. Only between 6 and 12 per cent of electricity is used to power the servers to perform computations; the rest is often used to keep them idle just in case there is a surge in activity.

According to cloud computing consultancy EMC and the International Data Corporation, in the United States in 2010, data centres used about 76 billion kWh, or about 2 per cent of all the electricity consumed in the country. This is increasing almost exponentially as the world gets used to video on demand and 3G and 4G telecommunications.

There are several issues here for energy efficiency:

1 The chips themselves are inefficient, producing lots of heat.
2 Just cooling them down uses a huge amount of power.
3 The large companies running them are rewarded for responding quickly to demands on server uses. They are not rewarded for saving energy; therefore they keep energy use at a maximum continuously.
4 With the inexorable expansion of 'cloud computing', this trend is set to continue.

For the energy manager of a data centre several solutions to these issues are possible. They are listed below.

Benchmarking

As with every form of energy management, knowing the true cost of energy used and how it is used is paramount. The metric for defining the efficiency of a data centre is its electrical power consumption. A ballpark figure for energy use of a

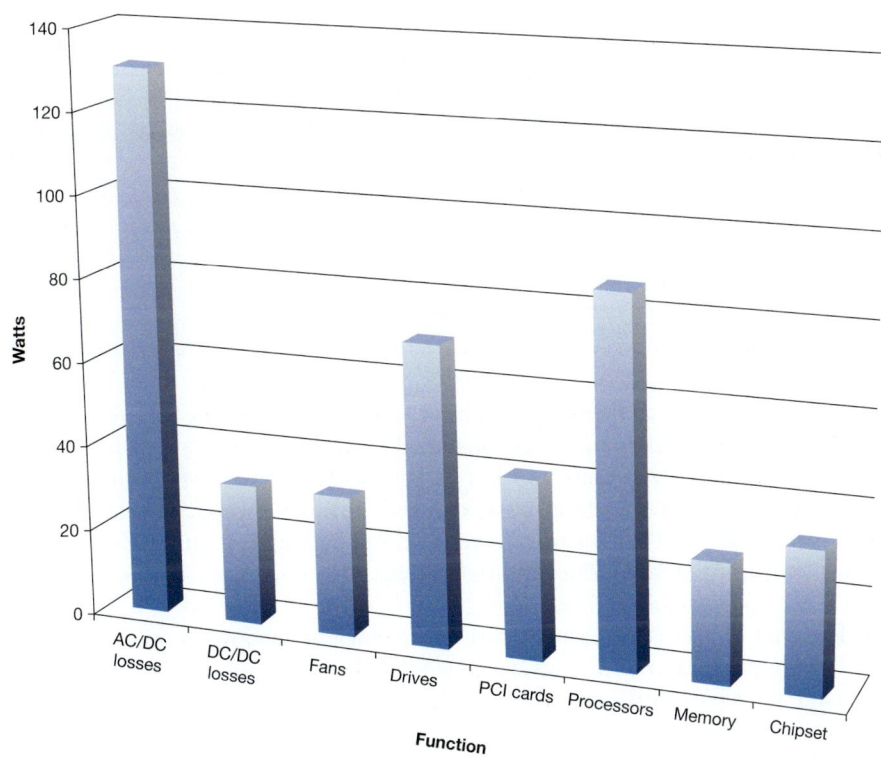

Figure 11.3 How a microchip uses power: based on a dual processor 450W 2U server, approximately 160W out of 450W (35%) are losses in the power conversion process. Each process emits heat, which represents wasted energy.

Source: Author

data centre would be $1,000 per year per kilowatt of IT load. This is based on a cost of $0.12 per kilowatt-hour. Over its 10-year life this would mean $10,000 per kilowatt of load. Typically around half the energy used by a data centre supports this IT load and the other half supplies its data centre physical infrastructure (DCPI), including the power supply. Thus this figure can be doubled. A 250kW data centre, over ten years, would face an energy bill of about $5 million.

The measure of the efficiency of a data centre is power usage effectiveness (PUE). This is the sum of all of the power consumed by the building (DCPI), including by the power distribution unit, UPS (uninterruptible power supply), lighting, chillers and so on, divided by that consumed by the IT equipment, which includes servers, memory backup and telecommunications equipment. This is as follows:

PUE = Total power used/Power used by IT equipment

The lower and closer this figure gets to 1, the more efficient it is. It is rarely possible to get it below 1.2. Improving the PUE from 2.0 to 1.6 for a data centre with a 2.5MW IT load yields energy savings of 20 per cent. Given a cost of energy of, say, $0.12/kilowatt-hour, this equates to annual savings of $1,200,000.

Terminology

PUE: power usage effectiveness
CRAH: computing room air handler
CRAC: computer room air conditioner
PDU: power distribution unit
UPS: uninterruptible power supply
DCPI: data centre's physical infrastructure

The first task then is to measure the energy consumption of the different pieces of equipment. Typically, energy is consumed as illustrated in Table 11.1.

Reducing the energy consumption of any of these pieces of kit produces a double win: not only are energy costs reduced, but power capacity-related costs are as well, since DCPI costs are reduced and less backup is required. Savings on energy consumption are reduced in two ways:

1 Through replacing equipment with more efficient versions.
2 Through operational savings, such as load shedding and draught sealing, to make the air conditioning more efficient.

Load shedding is the technique of powering off non-critical areas that are currently not required, whenever, say, the uninterruptible power supply (UPS) switches to battery, gets too hot, or the load reaches a certain set level. As a form of server power management it does reduce electricity costs but not necessarily the operational infrastructure electricity costs. High-efficiency servers

Table 11.1 Energy consumed

Equipment	Proportion of total energy consumed (%)
IT equipment	47 (PUE: 2.13)
Chiller	23
Humidifier	3
CRAC/CRAH	15
PDU	3
UPS	6
Lighting	2
Switchgear/generator	1

and high-efficiency UPS systems reduce both, and can result in savings more than four times that gained by power management alone.

Consolidate and update

Most data centres contain legacy technology platforms. Many of these, since they hold archived data, can be taken offline and powered down, even if not physically retired. Sometimes their data can be consolidated onto a fewer number of servers.

Most modern servers have power management features. It is important to check that this function has not been disabled and to enable it on all devices which have this capacity in order to reduce power consumption when computational load is reduced.

Figure 11.4 A blade server, or stripped-down server computer, in this case an IBM HS20 series.

Source: Wiki Commons (Robert Kloosterhuis)

When replacing servers, their performance should be compared using a 'per watt' metric. Most major suppliers of servers provide configuration tools that report actual power consumption for various configurations. This is even true with blade servers, which will not necessarily reduce power consumption unless used and installed appropriately. (A blade server is a stripped-down server computer with a modular design optimised to minimise the use of physical space and energy.) Sometimes optimisation may mean metering each individual server before estimating the power savings that could be gained by large-scale server migration.

As a rule of thumb, the following strategies can be effective:

- replacing two or more old servers with a single processor or a single two-way server;
- replacing an old server with a blade server based on a low-voltage or mid-voltage processor;
- if the server has dedicated disk drives, using class 2.5-inch drives instead of 3.5-inch drives;
- replacing dual processor servers with single dual core processor servers;
- replacing a four-way server with a two-way dual core server.

ENERGY STAR-labelled servers are available virtually everywhere in the world. Servers that meet ENERGY STAR energy efficiency requirements will, on average, be 30 per cent more efficient than standard servers and should obviously be chosen.

Soon, it will be possible to use micro-processing chips in servers that cost just 10 per cent of the price of currently used processors as well as consuming 10 per cent of the energy. These new '3-D Servers-on-a-Chip' are being developed by the EuroCloud server project, whose partners include Nokia, ARM and IMEC. This research project has adapted the low-power microprocessors found in mobile phones to work on a larger scale. Hundreds of microprocessor cores may then be embedded in a single server, making data centres containing one million cores feasible in the future. Because they are stacked in a three-dimensional matrix, any one part of the chip has greater access to another part, meaning shorter journeys for electrons, which means less energy lost and faster performance.

Optimise cooling

It is beneficial to group equipment with similar heat load densities and temperature requirements together. Racks should be arranged together in a hot aisle/cold aisle configuration and be isolated from each other. The cold air supply should be in the mid-70s Fahrenheit (around 23°C) and the hot air return could be as high as 90 to 100° Fahrenheit (over 33°C). That is why they should be kept separate.

Recommended and allowable airflow, filtration, humidity and temperature limits are all described in American Society of Heating, Refrigerating and Air-Conditioning Engineers (ASHRAE) publications such as *Thermal Guidelines for Data Centre Environments*. It should not be necessary to feel cold in a data centre. The recommended temperature is between 64.4°F (18°C) and 80.6°F (27°C). The

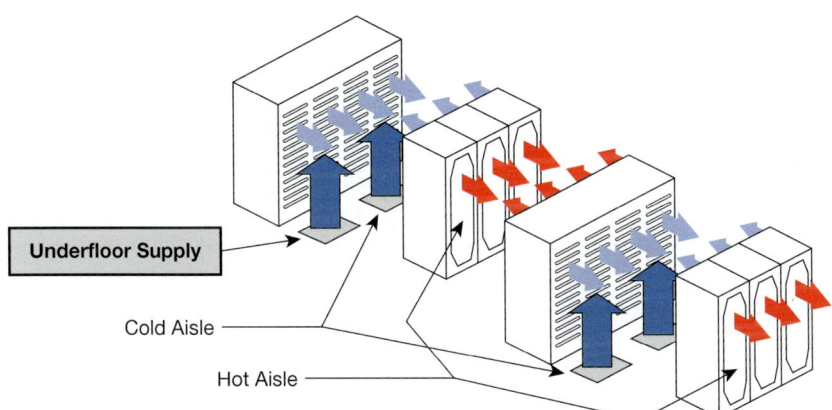

Figure 11.5 Optimal
cold and hot server
rack aisle layout.

Source: US Department of
Energy

use of medium-temperature chilled water (55°F (12.7°C) or higher) is acceptable. This improves plant efficiency and eliminates the need for the chiller during many hours of operation.

A central cooling plant and air handlers are usually more efficient than distributed air-conditioning units. Water is 3,500 times more effective than air on a volume basis for cooling servers. It is possible to purchase liquid-cooled racks. Manufacturers are also introducing liquid-cooled computers as well. An efficient water-cooled variable-speed chiller should be the first option, with high-efficiency air handlers and low-pressure-drop components. An integrated control system that minimises unnecessary dehumidification and simultaneous heating and cooling is also recommended.

It may also be possible to design the data centre for free cooling. On most nights and during mild winter conditions the use of outside air is perfectly possible; it is used by many data centres and is standard practice in the telecommunications industry. Nevertheless, a proper evaluation of the local climate conditions must be completed before opting for this measure. The use of variable-speed fans will greatly enhance efficiency as they can be matched with server flow requirements.

Virtualise

Virtualisation, or the simulation of the software and/or hardware upon which other software runs, enables several operating systems or processes to be run in parallel by different users on a single central processing unit (CPU). This tends to reduce overhead costs. It is part of the development of 'autonomic computing', which permits the server environment to manage itself based on perceived activity. The aim is to make more efficient use of server resources, to improve server availability, to assist in disaster recovery, testing and development, and to centralise server administration.

Eliminating a single server may permanently reduce the load by 200 to 400W, enabling a ten-year saving of $7,680 per server (at $0.12 per kilowatt-hour), which is substantially greater than the cost of the server itself. Some overhead is required to implement virtualisation, but this is minimal compared to the savings that can be achieved.

Standardise

Since it is often difficult to predict the performance requirement for server-based applications, managers often specify the highest-available performance, which also consumes the most energy. If servers are virtualised, the recommended energy-saving strategy is to use the highest-performance server to minimise overall power consumption. If the management policy is to deploy servers application by application, then the recommended strategy is to deploy them on the lower performance blade of a blade server by default, and only use a higher-performance blade if the need is demonstrated. This can save 10 per cent or more of structural IT load.

Rationalise memory storage

For storage devices, power consumption is approximately linear to the number of storage modules used. Therefore rationalising storage will reduce power consumption. This includes taking superfluous data offline. For that which cannot be taken offline upgrade from traditional methods to thin provisioning. Thin provisioning is the allocation of virtual RAM to running tasks as and when required, to make it seem as if the task has sufficient memory. This strategy is more efficient where the amount of resource used is much smaller than allocated so that the benefit of providing only the resource needed exceeds the cost of the virtualisation technology used. This maximises storage capacity use by drawing it from a common pool of purchased shared storage when needed.

Reduce DCPI consumption

Once the load demanded by the IT equipment is minimised and known, then it can be determined whether the data centre's physical infrastructure (DCPI) power supply, the power distribution units and/or transformers are over-specified. If they are running at below their optimum output capacity, they will be running inefficiently. This is the most common source of energy wastage. Power supply must be matched precisely to power usage.

Other sources of wasted energy include air conditioners. They should not be running with too low output temperatures, which in any case should not be lower than the exhaust temperature of the IT equipment. Nor should they be continuously dehumidifying air which must then be re-humidified by the humidifier. If they are forced to consume excessive power to drive air against high pressures over long distances they will also be operating inefficiently. There should be no heating going on while other units in the same room are cooling down. Cooling pumps should not have their flow rate adjusted by throttling valves as this dramatically reduces their efficiency. Instead, variable-speed drives should be attached to the pumps.

Most of the above problems are design problems. They can cause DCPI consumption to be twice what it needs to be. If the majority of these errors are found, it may be better to reconceptualise the whole design from scratch.

Starting from scratch

Whether redesigning an existing data centre or designing a brand-new one, there is plenty of help available to help them run efficiently and save money for their owners. What are called 'Total Cost of Ownership' analysis tools are available courtesy of the EuroCloud project. These will help designers rapidly assess the cost indications of different designs over the entire life span of the centre. A suite of benchmark software, available at http://parsa.epfl.ch/cloudsuite/, is available to analyse the common tasks of data servers: data serving, data analysis, media streaming, software testing, search and web serving.

If there is a choice of location of the data centre, then the coldest available place in the world should be selected. For example, BMW has just relocated its data centre requirements to a 45-acre complex in Iceland. If the data centre must be in a hot climate, then the exterior should be painted white and natural cooling deployed, together with solar-powered cooling (see Chapter 5).

Seal the server rooms

HVAC systems will not work properly unless the room they are cooling is sealed and made airtight. This is because replacing any cold air unintentionally lost is very expensive, but it is also a necessary condition of the fire code. This requires a proper inspection of the room. First, all the openings in the perimeter walls should be sealed. Cable trays and conduits that pass through them should be inspected for places where holes may not have been properly sealed. Other common places for air infiltration are entrance doors, windows and service entries. These must all be sealed or isolated.

Second, all openings in the raised floor should be sealed with proper materials such as foam, seals, fibreglass or grommets. If using tape, professional airtightness tape should be used. Third, bypass airflow within the cabinets must be prevented by installing tight-fitting blanking panels in unused rack openings. Special blanking panels are available for this purpose.

Then, and only then, as a fourth stage, the heat load would be modelled, taking account of the cooling capacity and airflow, to see how many cooling units need to be operating, how many perforated tiles need to be installed and where they should be placed. The heat load is determined by adding together all of the PDU or remote power panel outputs. Another way of doing this is to monitor the UPS outputs. If non-IT load is present in the room, subtract it when calculating the IT load on the raised floor. Optimise the cooling units by checking their temperature and humidity set point and sensitivities. Check the calibration of the return air sensors. Verify each cooling unit's actual delivered cooling capacity.

All of this should be carried out annually as part of a maintenance check, since it will change over the year. Recommissioning of servers and server racks should be done as a regular part of maintenance operations.

Underfloor and overhead obstructions often interfere with the distribution of cooling air, so check for these. A data centre should have a cable management strategy to minimise airflow obstructions caused by cables and wiring. A minimum effective (clear) height of 24 inches should be provided for raised-floor installations. Finally, continuously monitor temperature, humidity and underfloor

Figure11.6
Ready-made, airtight
and secure containers
for data centres
may be purchased,
complete with energy
management features.

Source: Alternate Energy
Source

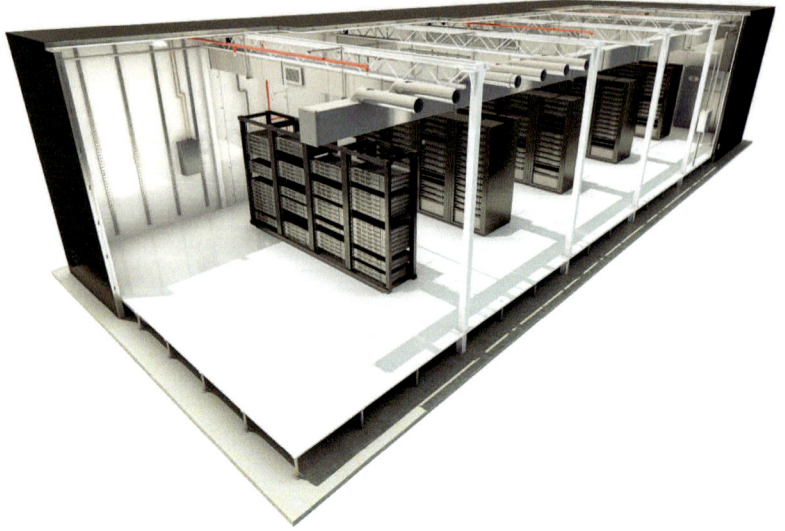

pressure. Ideally the software purchased should combine all of these data with power usage.

Use the surplus heat

Virtually all the energy consumed by a data centre ends up as waste heat. If a use can be found for this heat, for example, to sell it to nearby buildings, this would be a fantastic opportunity. The low-grade and variable nature of the heat suits loads that are on-site or adjacent, such as greenhouses, hospitals or swimming pools.

The average data centre has 50 times the power density per square meter of the modern office and 25 times that of a retirement home. Greenhouse heating demand is five times that of the average data centre. Depending on the respective sizes of a greenhouse and data centre, horticulture for a cash crop may well be an attractive option (see the case study at the end of this chapter).

Get people working together

IT and facilities staff must be encouraged to work together; otherwise they will suffer from 'conflicting prime directives' syndrome and be forever working at cross purposes: the former to prioritise fast response and downtime or failure minimisation, and the latter to optimise energy usage. If the energy manager cannot achieve this on their own, they should seek management cooperation.

Monitor in real time

Various tools are available to manage energy use, deploying historical data and real-time monitoring to project patterns of demands so that unnecessary backup is not kept running 100 per cent of the time. Operators then have the opportunity to shift power from cooling to IT whenever possible.

Real-time monitoring allows for the integration and correlation of power use, capacity/load levels and environmental information from throughout a single data centre or every data centre under management. The best software allows the facility manager to drill down to a single server in a single rack on a specific floor of a specific data centre to see how power is being consumed. Power capping, throttling, load shedding and shifting can be managed this way, leaving operators free to manually adjust capacity and load, and to respond to emergencies or events with planned actions.

Some solutions provide accurate power data at full-power levels, when powered off, when running idle, when fully loaded, and at peak. This means that racks of servers can be loaded safely to capacity, uncovering otherwise hidden capacity in the entire data centre. The advantage is that by allocating power based on actual consumption, computational capacity can be doubled for the same level of resources. Other advantages include the possibility of finding hidden storage capacity and prolonging operations during power and cooling outages. The energy consumption of standby infrastructure can also be drastically reduced.

In America, the Department of Energy data centre initiative, which includes the Industrial Technologies Program's Save Energy Now in Data Centres Program at the Lawrence Berkeley National Laboratory, develops tools and resources to make data centres more efficient.

Charge clients by usage

If a data centre has multiple clients, it is possible for data centre managers to charge clients separately by tracking their use in real time and billing them accordingly. Tracking power use and temperature on a per-customer basis allows for better planning.

Use active energy management measures

Newer generations of network equipment contain active energy management measures including idle state logic, gate count optimisation and buffer reduction. Ethernet network energy efficiency can be substantially improved by dynamically switching the speed of the network links to the amount of data that is currently transmitted.

In addition, a throttle-down drive is a device that reduces energy consumption on idle processors, so that when a server is running at its typical 20 per cent utilisation it is not drawing full power. The concern sometimes expressed by IT managers that this will compromise reliability is unfounded, since the hardware is designed to handle tens of thousands of on-off cycles. Server power draw can also be modulated by installing 'power cycler' software. This directs individual devices on the rack to power down during periods of low demand.

Improve power supply

Since server racks operate on direct current (DC), the provision of renewable energy will be far more efficient, since this is how it is supplied, and so there will be no losses of power in conversion to alternative current (AC). If there is no

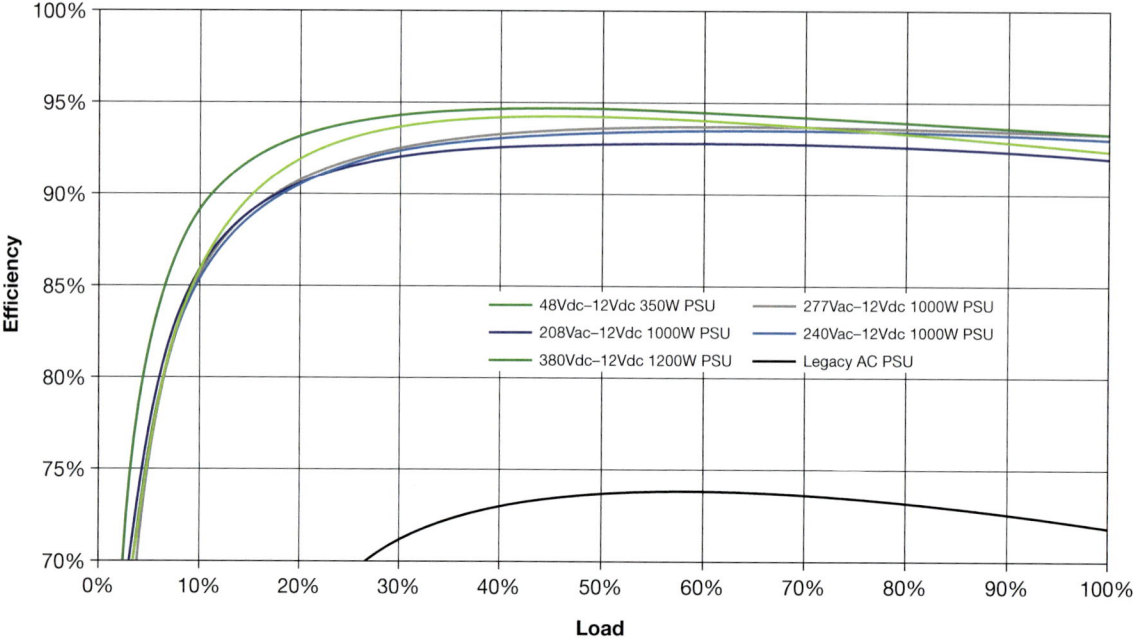

Figure 11.7 Power supply efficiencies at varying load levels for typical power supplies.

Source: Quantitative Efficiency Analysis of Power Distribution Configurations for Data Centres, The Green Grid

choice and AC power must be supplied, high-quality modern components and advanced engineering can convert from AC to DC with efficiencies of up to 95 per cent compared to traditional transformers' efficiency of 60 to 70 per cent. This also reduces cooling requirements.

To put this in financial terms, simply by increasing power supply efficiency from 75 per cent to 85 per cent, savings of up to $6,000 per year per rack are possible.

The provision of one power supply per server rack can increase the load factor to 70 per cent compared to individual server power supplies (20–25 per cent). Improvements in the efficiency of uninterruptible power supplies (UPS) can yield significant savings as well. Double conversion systems are those most commonly used in data centres. Their efficiency ranges from 86 per cent to 95 per cent. Upgrading this by just 5 per cent will reduce an energy bill by $90,000 a year for a 15,000 ft² data centre using 13,140MWh of energy per year for IT equipment.

Lighting

Occupancy sensors should be installed as standard to control lighting. All lighting should be either LED (by far the most efficient) or compact fluorescent.

Energy supply and storage

Data centres and telecommunications facilities are among those which require uninterruptible power supplies. It is not necessary for diesel generator backup to be running all the time in order to guarantee this. This section compares batteries, flywheels and super-capacitors, and their ability to provide rapid response to power spikes, cuts and lows. It also looks at using renewable energy and fuel cells to provide power.

The choice of backup is made on its efficiency, which is defined in exactly the same way as it is for the data centre as a whole: it is relative to the amount of energy required to keep the storage equipment charged. For a flywheel this would be the energy required to keep it spinning, called the standby loss. For batteries it is the energy required to maintain its float charge, called the trickle charge loss. This is energy which is lost forever.

By this comparison, batteries would be more efficient. Losses for a flywheel are between 0.2 per cent and 2 per cent of its full rating. Losses for a battery are around 0.2 per cent. Over its lifetime, therefore, an inefficient flywheel could cost the owner of a 1MW data centre $306,000 more than using a battery. However, by integrating the flywheel into a UPS, which shares their control and fan power, and using several high-efficiency flywheels in series, these losses can be reduced to 0.2 per cent to 0.5 per cent.

But the total lifetime cost of flywheels may be cheaper than batteries, since they will last longer in cases where there are frequent outages of short duration,

Figure 11.8 A bank of batteries used as an uninterruptible power source (UPS) in a data centre.

Source: Thinkstock

as this reduces the lifespan of lead acid batteries. Batteries are also heavier; super-capacitors and flywheels will be lighter, and this may be a consideration. Batteries are also more polluting at the end of their life, especially NiCad batteries. They must be disposed of properly.

Batteries usually provide 5 to 15 minutes of backup power, while a flywheel can typically supply from 8 to 15 seconds of backup at full power. But if flywheels are used with a battery-based UPS, they are capable of handling all short-duration power disturbances, leaving the batteries only for extended outages and thereby extending battery life. A flywheel can also work with a rotary UPS (which contains a flywheel) in conjunction with a generator and act as an alternative to a battery-based UPS.

Rechargeable batteries

- *Nickel Cadmium (NiCad):* can be recharged many hundreds of times, though they have a 'memory' that can compromise life span. Cadmium is toxic and poses an environmental hazard. Almost completely phased out and should never be used. Old batteries should be disposed of carefully.
- *Nickel Metal Hydride (NiMH):* does not suffer from the same memory effect as NiCad. Increased energy density (power/weight) offers longer use between charges. Lack of cadmium makes NiMH more environmentally friendly.
- *Lithium Ion (LiI):* has higher energy density. Often found in laptop computers and cell phones. One-third the size and one-fifth the weight of a lead acid battery. Does not suffer from memory issues or contain heavy metals such as cadmium or mercury.
- *Lead acid:* the most widely used battery for telephone and data centres, grid energy storage, uninterruptible power supplies and off-grid electric systems because of their high power density and ease of use. They can withstand frequent discharging. Valve-regulated lead acid (VRLA) batteries are sealed, and therefore the most common, since they do not require special ventilation, as do 'flooded' or 'wet cell' lead acid batteries, which vent hydrogen. Vented batteries are usually installed on open racks in dedicated battery rooms that

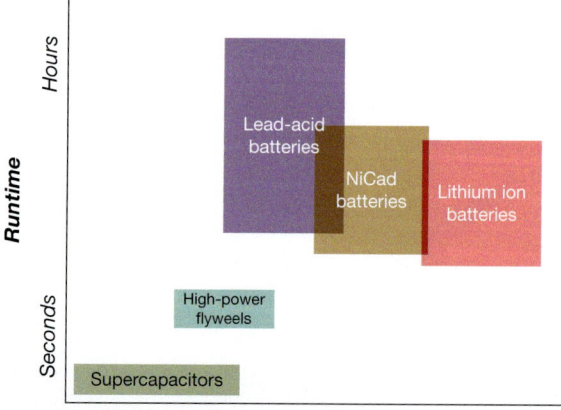

Figure11.9 The relative cost versus running time of different energy storage equipment. The further to the right, the more expensive per kilowatt; the further up, the longer they can provide power. Although super-capacitors and flywheels are cheaper than batteries, batteries, especially lead acid batteries, will provide power for longer.

Source: Author

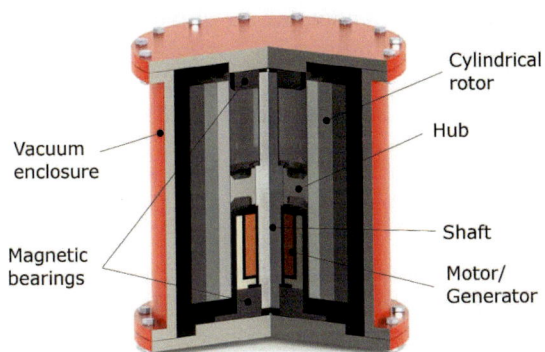

Figure 11.10 Cross section through a flywheel used for energy storage.

Source: Wikimedia Commons (Pjrensburg)

Figure 11.11 A flywheel at a trade show.

Source: Wikimedia Commons (unattributed)

have separate ventilation systems to prevent the contaminated air from mixing with that for the rest of the building. VRLA batteries may be stored anywhere. Battery recycling is widespread.

Flywheels

Modern super-flywheels store energy mechanically and can release it very quickly: the perimeter of a flywheel can travel at twice the speed of sound. Because of this, they may require more maintenance over their lifetime than a battery. They can be recharged very rapidly as soon as they have discharged. They come in rated capacities of 0.4–20MW, possess cycle efficiencies of 80–95 per cent, and cost a little more than lithium batteries, but this is compensated for by longevity; they can last up to 20 years.

Advantages

- fast recharge after use;
- makes more economic sense for applications of 500kW or above;
- wide operating temperature range (0° to 40°C (32° to 104°F)) compared to batteries;
- lifetime of more than 15 years;
- have a power density advantage over batteries when less than a minute of running time is acceptable;
- in larger applications, may have smaller footprint than batteries (e.g. > 50kW).

Disadvantages

- maintenance cost;
- short running time translates into longer generator running times (noise, fuel consumption, pollution);

- flywheels function with the assistance of a complex set of multiple controls which represent potential single points of failure;
- complexity of installation;
- efficiency losses to maintain flywheel rotation (during normal operation);
- can have a larger footprint than batteries in small applications (e.g. <50kW).

Super-capacitors

Whereas batteries store energy in chemical form, capacitors store it by amassing electrical charge on two electrodes. The larger the electrodes and the closer they are, the more energy may be stored.

Like flywheels, super-capacitors can be charged and discharged in seconds and can withstand many hundreds of thousands of such charging cycles. This is ideal for energy-saving applications that capitalise on transient opportunities for recharging. The main producers of super-capacitors worldwide are Nesscap Energy based in South Korea, and the California-based company Maxwell Technologies. Expensive now, they will soon come down to affordable levels, especially with regard to life-cycle costs.

Advantages

- can process a large number of charge and discharge cycles without suffering wear;
- satisfactory operation in a wide range of temperatures, unlike batteries;
- its energy density is 20 per cent of the battery's, making it relatively small in size and weight for short discharge times (seconds).

Disadvantages

- high cost for longer running times (minutes to hours);
- can only store low quantities of energy (short running times);
- very short history in data centre environments (less than ten years) – no extended performance data.

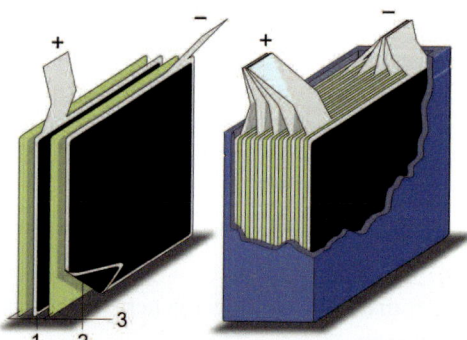

Figure 11.12 Cross section through an electric double-layer capacitor (activated carbon electrode, box type), a type of super-capacitor. *Key:* 1. polarisation electrode with collector (positive pole); 2. Polarisation electrode with collector (negative pole); 3. Separator.

Source: Wikimedia Commons (Tosaka)

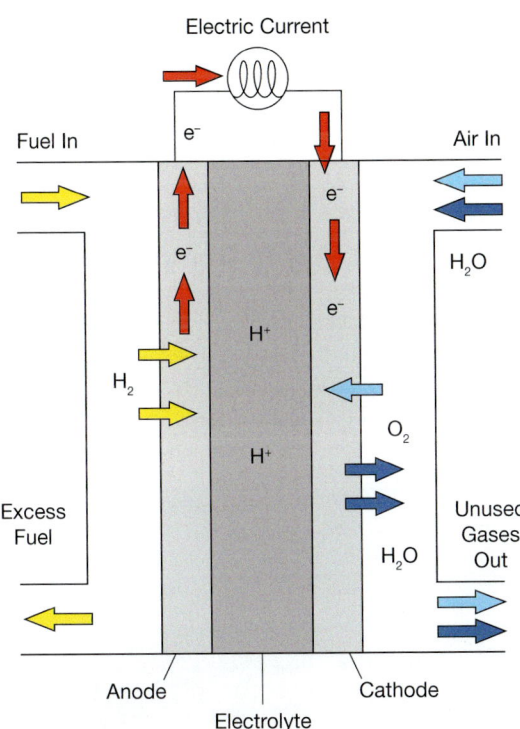

Figure 11.13 Schematic diagram of a proton-conducting fuel cell. Once charged with fuel, typically hydrogen or methanol, it can generate electricity by reaction with air, discharging only water.

Source: Wikimedia Commons (R.Dervisoglu)

Fuel cells

A fuel cell converts the stored chemical energy in a fuel into electricity through a chemical reaction with oxygen or another oxidising agent. Fuel cells are an alternative to diesel generator backup. They are environmentally and climate friendly if the fuel to power them is renewably sourced. Typically this fuel is hydrogen, but some fuel cells use methane or methanol. Hydrogen can be generated on-site from renewable energy such as wind power or solar power, or from grid-sourced power. The methane can be fossil fuel, supplied by the grid, or renewable, supplied as a treated biogas from anaerobic digestion.

The gas is stored in tanks for use when required. Fuel cells can run indefinitely as long as fuel is supplied. Therefore, the more tank capacity, the longer they can run. If the facility generates its own renewable electricity, hydrogen is an extremely valuable means of storing this energy at times when it cannot be used while being generated.

Advantages

- clean, with no hazardous materials;
- silent and vibration-free;
- lightweight and compact;
- few moving parts;
- long-lasting.

Disadvantages

- requires extensive site preparation;
- ideally requires on-site renewable electricity generation capacity;
- cost of processing and storing the hydrogen or methane fuel.

Case study: Apple Computers, USA

Computer giant Apple has installed 10MW of fuel cells running on biogas from a landfill site at its data centre in North Carolina. They are supplied by Bloom Energy. In addition, there is an array of 20MW of solar PV. All of this renewable electricity is being exported to the local utility Duke Energy, as well as being used on-site.

Figure 11.14 Apple's data centre in North Carolina.

Source: Apple Computers

Conclusion

Table 11.2 gives a summary of some of the measures described above and the relative savings that may be achieved.

Clarifying costs and benefits for management

Senior management is more likely to be brought onside the energy efficiency agenda by being shown a comparison of the total life-cycle cost of ownership of an installation with and without power management and optimisation for maximum efficiency.

Energy consumption requires monitoring in very specific ways, and the benefits of different measures require complex calculations. They are not as straightforward as in other cases, as we have shown. Data from Atrium Computing and the Statistical Office of the European Commission (EUROSTAT) in 2010 show the following savings that are possible for a data centre with four pods of 1MW (IT equipment power), which a reduction of only 0.1 percentage points of the PUE would save after ten years (see Table 11.3).

Table 11.2 Summary of energy efficiency measures that may be applied to data centres

Measure	Savings	Guidance	Comments
Correctly size DCPI	10–30%	• Use a modular, scalable power and cooling architecture	• For new designs and some expansions
Virtualise servers	10–40%	• Consolidate applications on to fewer servers • Liberate power and cooling capacity	• Major operation • In an existing facility turn off redundant power and cooling equipment
More efficient air-conditioner architecture	7–15%	• Target cooling properly: shorter air paths require less fan power • CRAC supply and return temperatures are higher, increasing efficiency, capacity, removing need for dehumidification	• For new designs • Benefits are limited to high-density designs
Use economiser modes on air conditioners	4–15%	• Savings depend on geographic location • Check economiser function is not disabled • Employ variable speed fans	• For new designs • Difficult to retrofit
More efficient floor layout	5–12%	• Implement cold and hot server rack aisle layout	• For new designs • Difficult to retrofit
More efficient power equipment	4–10%	• Latest high-efficiency UPS systems have 70% less losses at typical loads • Light-load efficiency is the key parameter, *not* the full-load efficiency • UPS losses must be cooled, doubling their costs	• For new designs or retrofits
Coordinate air conditioners	0–10%	• Ensure cooling and conditioning units are working in harmony and not against each other • May require a professional assessment to diagnose	• For any data with multiple air conditioners
Locate vented floor tiles correctly	1–6%	• Position vented tiles correctly and ensure the correct number are installed • Professional assessment may be required	• Only for data centres using a raised floor • Easy, but requires expert guidance to achieve best results
Install energy-efficient lighting	1–3%	• Use occupancy sensors • Use LED lighting technology	• All data centres can benefit
Install blanking panels	1–2%	• Decrease server inlet temperature • Also saves on energy by increasing the CRAC return air temperature • Cheap and easy with new snap-in blanking panels	• For any data centre

Table 11.3 Estimated financial savings after 10 years in four European countries for a data centre with four pods of 1MW (IT equipment power) with a reduction of just 0.1 percentage points in Power Usage Effectiveness (PUE). Source: Atrium Computing and the Statistical Office of the European Commission (EUROSTAT), 2010

Country	Saving
France	€1,955,232
Ireland	€3,239,448
Netherlands	€3,018,696
United Kingdom	€2,994,168

This is a significant amount of money.

Case study: the Condorcet facility in Paris, France

Figure 11.15 Condorcet data centre, Paris, France.

Source: European Union Joint Research Commission

This was the first and only data centre operator to adopt the EU Code of Conduct for Data Centres across all of its sites. Built on a former industrial site, this groundbreaking building uses only natural materials in its construction. This means an absence of PVC for ducting and so on.

Advanced filtration technologies remove the need for chemicals in air conditioning. The climate control system uses rainwater collected from the roof and stored in buffer tanks. Designed in a modular fashion to maximise efficiency, it naturally draws in external air to provide free cooling. The roof is also painted white to reduce the amount of cooling needed. Effective energy distribution limits the amount of technical equipment and gives an overall reduction of 25 per cent on electrical installations of equivalent power and performance. The rooms are fed by a three-way power supply network to optimise equipment use, providing uninterruptible power, with hydrogen fuel cells used as backup. As for lighting, no incandescent or halogen light bulbs are used.

Excess heat is channelled into a semi-tropical arboretum in a greenhouse that is integrated within the building, and which is used by The French National Institute for Agricultural Research (INRA) to grow and research species most adaptable to changes in the prevailing climatic conditions in France.

Further information

The Certified Energy Efficiency Data Centre Award (CEEDA) from the British Computing Society provides data centres with an independent means of assessing and accrediting their facilities for energy efficiency. It enables organisations to meet the needs of increasing carbon legislation, reduce energy costs, and benchmarks them as leaders in data centre energy efficiency. See http://ceeda.bcs.org/category/17062.

The European Union Joint Research Commission (EU-JRC) runs a voluntary Code of Conduct on Data Centre Energy Efficiency. It is being applied globally, and is self-assessed. See http://re.jrc.ec.europa.eu/energyefficiency/html/standby_initiative_data_centres.htm and http://iet.jrc.ec.europa.eu/energyefficiency/ict-codes-conduct/data-centres-energy-efficiency.

Green Touch, a consortium of leading information and communications technology (ICT) experts dedicated to fundamentally transforming communications and data networks and increasing network energy efficiency by a factor of 1,000 from current levels. See www.greentouch.org.

FIT4Green: Federated IT for a sustainable environmental impact.
See www.fit4green.eu.

GAMES: Green Active Management of Energy in IT service centres.
See www.green-datacentres.eu.

All4Green: Active collaboration in data centre ecosystem to reduce energy consumption and GHG emissions.
See www.all4green-project.eu.

CoolEmAll: Platform for optimising the design and operation of modular configurable IT infrastructures and facilities with resource-efficient cooling.
See www.coolemall.eu.

Notes

1   http://www.baaqmd.gov/Divisions/Engineering/Air-Toxics/Toxic-Air-Contaminant-Control-Program-Annual-Report.aspx (accessed 7 February 2013).
2   http://www.mckinseyquarterly.com/Data_centres_How_to_cut_carbon_emissions_and_costs_2255 (accessed 18 January 2013).

12
Minimising water use

No resource is more fundamental than water to the health and security of people and the environment. Failure to act risks exposing organisations to water scarcity issues, which could lead to dramatically increased costs, or even bring operations to a standstill. Currently, organisations waste a huge amount of water. The subject is often even further below the horizon of management boards than energy saving. Yet economical use of water can result in huge financial, energy and carbon savings, for example, in reduced pumping costs. Only one in seven senior executives of large companies in the UK, USA, China, South Korea and Brazil had set water reduction targets or publicly reported on water performance according to a recent survey.[1]

But legislation is looming. There is a global water crisis due to competition for water between agriculture, industry and people. Through rigorous, independent certification, businesses can demonstrate their journey towards improved water stewardship. Initiatives such as the UK Carbon Trust Water Standard fit exactly this purpose. Organisations who have already achieved the standard include Coca-Cola Enterprises, Sunlight PLC, Branston and Sainsbury's.

Case study: Sainsbury's, England

The Head of Sustainability, Engineering, Energy and Environment at Sainsbury's, Paul Crewe, has achieved a target of a 50 per cent relative reduction in water use throughout the supermarket chain, which is a saving equivalent to 393 Olympic-sized swimming pools each year. Strategies have included eradicating underground leaks, fitting pre-rinse spray taps and low-flush toilets in all their stores and investing in rainwater harvesting for all new stores as standard, as well as retrofitting these units in existing stores. Another innovation is reclaiming water from car washes. This has resulted in savings for individual stores of hundreds of thousands of pounds each year. Using rainwater for toilet flushing in one city store (as measured in Swansea, Wales) achieves an annual mains water consumption saving of 1,300m³ (286,000 imperial gallons).

Saving water also saves carbon emissions. In the UK, mains water contains an average of 1.47kg of embodied carbon dioxide equivalent per 100 litres, through

Table 12.1 Average efficiency of different measures

Efficiency measure	Percentage of water saved
Closed loop recycling	90
Closed loop recycling with treatment	60
Automatic shut-off	15
Counter-current rinsing	40
Spray/jet upgrades	20
Reuse of wash water	50
Scrapers	30
Cleaning-in-place (CiP)	60
Pressure reduction	Variable
Cooling tower heat load reduction	Variable

sourcing, transportation and heating. The embodied carbon contained in water supplies elsewhere will depend on the proportion of fossil fuels used in grid-supplied electricity in the local area.

Industrial uses of water are manifold. Examples include steam used during processing, steam cleaning, in evaporative cooling towers, and for the abatement of processing odour via wet scrubbers. Minimisation opportunities exist in almost all industrial plants. Potential savings are hard to estimate as they are process specific, but they can be as high as 90 per cent. Table 12.1 shows the average that might be saved for different types of efficiency measures.

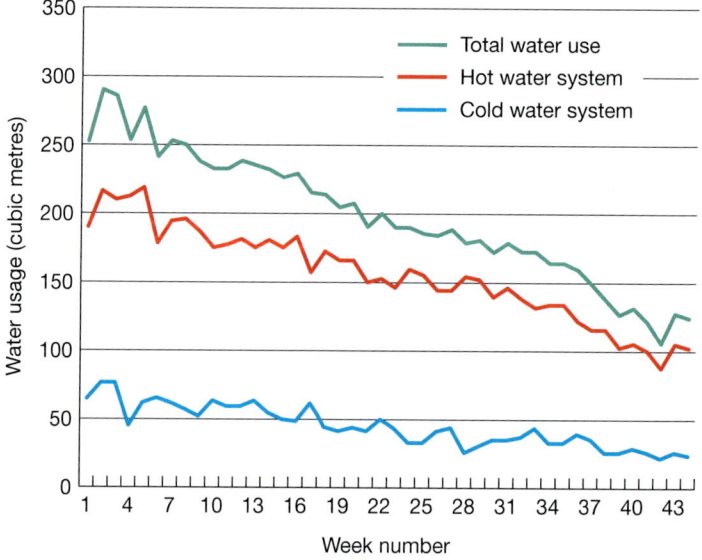

Figure 12.1 Water reduction experienced by a food company through using hoses fitted with nozzles and controls for cleaning purposes.

Source: WRAP

Conducting an audit

Just as with energy, an audit of water use is the place to start. This will help to identify very simple ways to easily capture and reuse water without extensive capital investment. The audit should cover all the sources of water use. It will consist of a list in spreadsheet form of water-using outlets, their function, flow rate and usage per week or day. Comparing this to utility bills or readings from a building management system can reveal discrepancies and targets for action. The audit should also quantify the true cost of water to the organisation. The end result should be a table with headings similar to the following:

Item	Location	Number of units (A)	Flow rate (gallons or litres/minute) (B)	Operating time (minutes/day) or (uses/day) for each unit (C)	Water used (gallons or litres/day) = A × B × C	Comments

The penultimate column may then be added up to determine the weekly/daily/ monthly total usage. While doing this, it is a good opportunity to discuss with staff members how they use water and ways to minimise use.

Water use can be checked against bills. Bills identify water use by volume. The average office water use is 4 cubic metres per employee a year.[2] This enables water use per employee, in an office, to be benchmarked by dividing annual water use (in volume) by the number of staff (full- and part-time). For example, for an office with 36 full- and part-time staff and a six-monthly water use of $101m^3$ (taken from the bill) this would be:

$101 \times 2/36 = 5.61m^3$/employee/year. This signifies that there is room for at least 20 per cent savings.

A water efficiency strategy

The next stage is a water efficiency policy and strategy. Most organisations already have a water management strategy; if so, it should be reviewed annually. As with electricity and energy use, the single most important aspect of water use is to minimise it. If the demand is minimised, the energy and financial cost of supplying it is reduced.

A good water use policy should:

- set a target for reducing water use;
- embed the targets within organisational policy and processes;
- set corresponding requirements in project procurement and the supply chain;
- measure performance at a site level relative to a corporate baseline, and report annually on overall corporate performance.

The units or key performance indicators (KPIs) chosen may be any of the following:

- volume of potable water consumed: gallons or m^3 per year, with reference to a unit of production[3]; or
- square feet or m^2 Net Lettable Area per year; or
- % reduction in use relative to 20xx.

Minimising water use

The Water Minimisation Hierarchy in a nutshell consists of the following priorities, in order:

1 eliminate wasted water;
2 improve efficiency and use alternative sources;
3 reuse water;
4 recycle water.

Initial steps to minimise use may be obtained by:

- having an organisational water efficiency policy;
- installing metering and sub-metering, perhaps in conjunction with a Building Energy Management System;
- installing water-efficient components;
- replacing potable water with water from other sources (e.g. rainwater or greywater) where appropriate;
- minimising the energy and carbon emissions associated with the generation, storage and supply of hot water (within the property);
- influencing user behaviour through water system design;
- regular checking and maintenance to ensure that water-consuming fittings, appliances, controls, pressure/flow regulation and monitoring systems are adequately maintained and work safely and in line with their design performance;
- having a targeted replacement or retrofitting programme for fittings or appliances.

Water meters alone very often have the effect of reducing use by up to 17 per cent. The energy manager should prepare a financial case for action based on the estimated capital and life-cycle costs of any work required and the value of reduced water and energy use. Consideration should be given to the timing of any investment so that it matches existing plans for expenditure. As with an energy efficiency action plan, a water efficiency plan may address the following factors:

- The project's current or projected installation of water-consuming fittings and appliances and their current or likely usage.
- The objective (e.g. maximum consumption level, or corporate target for improvement).

- Forecasts of alternative outcomes for end-use water consumption arising from the use of components with different levels of water efficiency (in the same unit as the chief metric).
- A site-specific target for water use (design intent or in-use), and/or specifications for different types of fitting and appliance.
- Projected financial, water and energy savings and associated financial costs.
- A log of designed-in and actual water use (recorded over time), supported by evidence of actions taken.
- A procedure for monitoring and review of performance against the target, together with a timetable for updating the water efficiency plan and capturing lessons learned.
- Allocation of who is responsible for implementing the actions.

Checking for leaks

Checking for leaks should be done regularly. One way to do this is to take a meter reading last thing at night and again first thing in the morning, or when it is not expected that any water should be used. If the reading has changed it may be due to a leak. Daily or weekly logging of water meter readings and working out how much average water is used for each purpose can yield a benchmark from which to improve. Washers in dripping taps and faulty valves should be replaced immediately. Leak-monitoring equipment should be installed if possible and auto shut-off of flow to unoccupied areas.

Figure 12.2 Leaks can waste a fortune in water.

Source: WRAP

Comparing the results of meter readings to water bills can reveal discrepancies, such as more water being billed for than is actually metered. This means that a thorough check should be undertaken. A leak detection subcontractor might be called in, as leaks can take place underground.

Taps

A running tap can waste over 6 litres of water per minute. Spray taps can save about 80 per cent of water used by conventional taps. A 'Tapmagic' insert can be fitted to most taps with a round outlet hole or standard metric thread. At low flows, Tapmagic delivers a spray suitable for washing hands. As the flow is increased it opens to allow full flow. Tap controls are available in both new and retrofit versions, including infrared (PIR), battery operated and simple push-top and spray taps. PIR sensors detect when a hand is beneath and turn on automatically. Basin taps should be limited to 4 to 6 litres per minute and sink taps to 6 to 8 litres per minute.

Toilets

Installing urinal flush controls typically saves around 70 per cent of water used for flushing. One organisation saved £1,300 ($1,800)/year in water and sewerage costs by installing a passive infrared (PIR) sensor. With an installation cost of around £960 ($1,300), a payback of just over eight months was achieved. Restrictor valves can be easily fitted to supply pipes to keep the water flow constant, regardless of fluctuations in water pressure.

Hot water systems

Sites using mains-pressure hot water will tend to use more water than those with a gravity system due to the higher flow rates. However, with good design, efficiency savings can be achieved with mains pressure hot water by fitting:

Table 12.2 High-efficiency target levels for water use

Fitting/appliance	Ideal water use target (at pressures up to 5 bar)
Shower (mixer) ≤	6 l/min
Shower (electric) ≤	6 l/min
WC ≤	3.5 l/flush (single flush or effective flush)
Urinal ≤	3 l/bowl/hour during building occupancy with user-presence activated flush, 0 l/hour outside of occupancy and activation period, with minimal water use in maintenance. Or fit waterless urinals
Tap ≤	6 l/min

Source: WRAP

Table 8.2 Dead leg volumes

Pipe diameter	10mm	15mm	15mm	22mm	22mm
	plastic	plastic	copper	plastic	copper
Litres per 10m pipe run	0.6	1.1	1.5	2.4	3.1
Max length for 1.5 litre dead leg (m)	25.0	13.0	10.0	6.0	5.0
Length for 30-second wait with 1.7 litres per minute spray fitting (m)	14.0	8.0	6.0	3.5	3.0

Figure 12.3 An aerated tap saves water but gives the impression of a full flow of water.

Source: Wikimedia Commons (Nikthestoned)

- small bore pipes (10mm is the smallest available);
- regulated aerators;
- low-water-use outlets;
- pressure and flow regulators.

If the mains or header tank pressure is greater than 3.5 bar pressure, regulators should be installed to maintain a constant pressure, to limit pressures in mains-fed hot water systems and to prevent damage to fittings. These regulators are normally adjustable and maintain a constant flow independent of supply pressure and can be fitted in-pipe or at each tap or shower. An alternative is a tap outlet with a built-in regulator.

All hot water pipes should be properly insulated and sited above cold water pipes to reduce heat transfer. A radial layout for pipes to outlets from the tank will also help keep heat losses down. The bore (diameter) of a pipe has a great impact on energy wastage, as may be seen in Table 12.3. The narrowest available bore should be chosen, and dead legs (the distance travelled between the boiler and tap) minimised to avoid losing heat.

Figure 12.4 Properly insulated boilers.

Source: Royalty Free Picture Agency

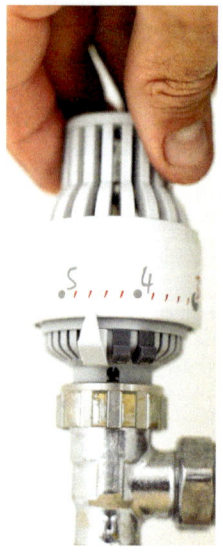

Figure 12.5 Using thermostatic control of individual radiators to make sure heat is only supplied where needed and to the correct level.

Source: Royalty Free Picture Agency

Figure 12.6 Well-insulated hot water pipes.

Source: Royalty Free Thinkstock

Reducing process losses

Pumps

Motor-driven systems account for approximately 40 per cent of global electricity consumption, with pump systems being a major component. Efficiency is governed by the ISO 14414 Standard for Energy Efficiency Testing of Pumping Systems.

A pump will operate efficiently when installed in a poorly designed or mismatched system. This is defined by the H/Q performance curve, which shows the relation between flow and head, where Q (flow) is the x-axis and H (head) is the y-axis.

The units for Q are normally [m³/h] or [l/s]. The unit for H is normally [m].

H (head) can be recalculated to p (pressure) by using the following equation:

$$p = Ú \times g \times H \text{ [pa]}$$

Where:

p = pressure [pa]
Ú = density [kg/m³]
g = acceleration due to gravity [m/s²]
H = head [m]

Figure 12.7 shows a pump performance curve and its best efficiency point (BEP). This is the unique point at which the pump operates at peak efficiency. The operational point should be within 85 and 105 per cent of this point. The figure also shows a system curve which defines the relationship between the head and flow through the system. It is defined by the intersection of the pump performance and the system curves. Matching these will ensure operation at or close to the BEP.

Pump efficiency can deteriorate, especially when liquids contain solid content. The veins should therefore be inspected and cleaned regularly. Several methodologies exist to determine system efficiency. The basic principle is to improve the volume of liquid delivered compared to the energy supplied to the pump drives. It can be calculated from volume and energy data collected over a period of time, or from instantaneous flow and power data. Pumping system efficiency (Ësys) is defined as follows:

$$Ësys = (Qreq \times Hreq \times SG)/5308 \times Pe$$

Where:

Qreq = required fluid flow rate, in gallons per minute
Hreq = required pump head, in feet
SG = specific gravity
Pe = electrical power input

Only the required head and flow rates are considered in calculating system efficiency. Unnecessary head losses are deducted from the pump head, and unnecessary bypass or recirculation flow is deducted from the pump flow rate.

Potential energy savings may be determined by using the difference between the actual system operating efficiency and the design (or optimal) operating efficiency, or by consulting published pump curves for design efficiency ratings.

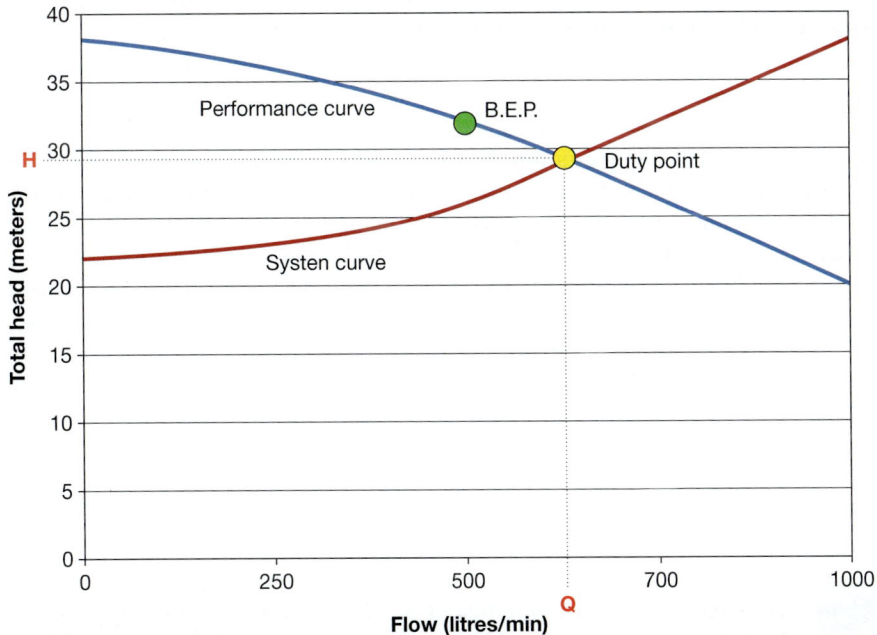

Figure 12.7 A pump performance curve, its best efficiency point and duty point.

Source: Author/Bob Went

Cooling water

Most process operations use cooling water, which can be collected and recycled. Often cooling towers are used to cool the water; where refrigeration chillers are used they should be replaced, as they are large users of energy. Minimising the flow of cooling water and by not cooling to a temperature lower than is actually required are other ways of saving energy. Reducing the bore of pipes can save a considerable amount of water, as well as meaning that pumps have to do less work, up to a point at which friction losses reverse this trend. Pumps should have variable-speed drives attached to them whenever a fixed flow is not required.

Boilers

The optimisation of energy use in boilers is covered in Chapter 5, under boilers and heat reclamation.

Steam

Many of the remarks for water apply to steam, including leak detection, use of reduced bore piping and variable-speed drives for pumps. Insulation should be continuous, sealed and as high as possible. Any redundant areas of steam delivery must be zoned off. Steam traps must work effectively. Although some steam will always condense in a distribution network, removing condensate without losing too much steam is a great way of improving efficiency. Yet much steam is lost through 'sticking' steam traps, so they should be checked regularly as part of a maintenance survey. Moreover, steam condensate contains valuable heat which can be captured and returned to the boiler.

Cleaning

Nozzles used for cleaning equipment, etc. may be improved in a similar fashion to taps. A leading drinks manufacturer saved £6,000 a year by using water-saving nozzles with controls on its container-cleaning equipment. The devices paid for themselves in less than ten weeks. Hoses should have a narrow bore and trigger nozzles may be colour-coded to ensure wastewater is discharged to the correct drain to prevent contamination. For cleaning purposes, dry-cleaning methods may be adopted that keep water use to a minimum. Often costs can be recouped in under a year by reduction in water and trade of room costs.

Figure 12.8 Spray nozzles in action.
Source: WRAP

Distillation

Distillation should not be carried out to such a degree that the products are being purified to a greater extent than necessary. Often savings can be made by using more efficient distillation trays or column packing. Distillation columns should be configured in the optimum way and heat reclaimed where possible. In some cases there are less energy-intensive alternatives to distillation, such as crystallisation and pervaporation.

Evaporation

Evaporation removes water and concentrates a liquid. Minimising the amount of water added to the feed liquid upstream is the chief way of reducing energy loads. Most other optimisation strategies involve redesign of the process and recovering the heat of the evaporated water, reusing the water, and the use of a mechanical vapour compression to deliver heat to the operator. Multiple-stage evaporation can save energy by letting each successive stage be at a lower pressure than the previous stage, and using the steam from the preceding stage as a heat source.

Drying

Again, the energy required for drying can be saved by minimising the water content of the solid in the first place by redesigning the process, or by using a mechanical means of removing much of the water prior to the drying stage. The drying area should be insulated and the evaporated water, if possible, condensed and reused. Some dryers can be optimised by adjusting their air input and airflow patterns. There may also be potential for recovering waste heat.

Humidity control

Some processes require a certain level of humidity. As this is very energy intensive it should be checked whether it is essential. If it is, then the minimum and maximum humidity levels should be kept as respectively low and high as possible to minimise energy use.

Reuse

Commonly, wastewater may be treated by an existing effluent treatment plant, using a dissolved air flotation (DAF) unit to clarify the wastewater, followed by a submerged biological filter, prior to discharge to sewer. Simply by investing in membrane technology (micro-filtration and reverse osmosis plant) to further treat the effluent, the quality of the treated water is such that it may be reused on-site for certain applications, such as boiler feed water, top-up feed for cooling towers, washing non-sensitive items, etc. The treated water offers a high level of purification and a relatively low energy and carbon footprint. It is stored in a water reuse tank prior to use.

The stages of this complete process are as follows:

1   filtration to remove coarse solids;
2   pumping into the bioreactor where the majority of the chemical oxygen demand (COD) is removed;
3   treatment by ultrafiltration to remove biomass;
4   filtration using reverse osmosis.

Overall, this treatment removes over 99 per cent COD and suspended solids loads. Typical payback period for the installation of a complete system is between two and three years.

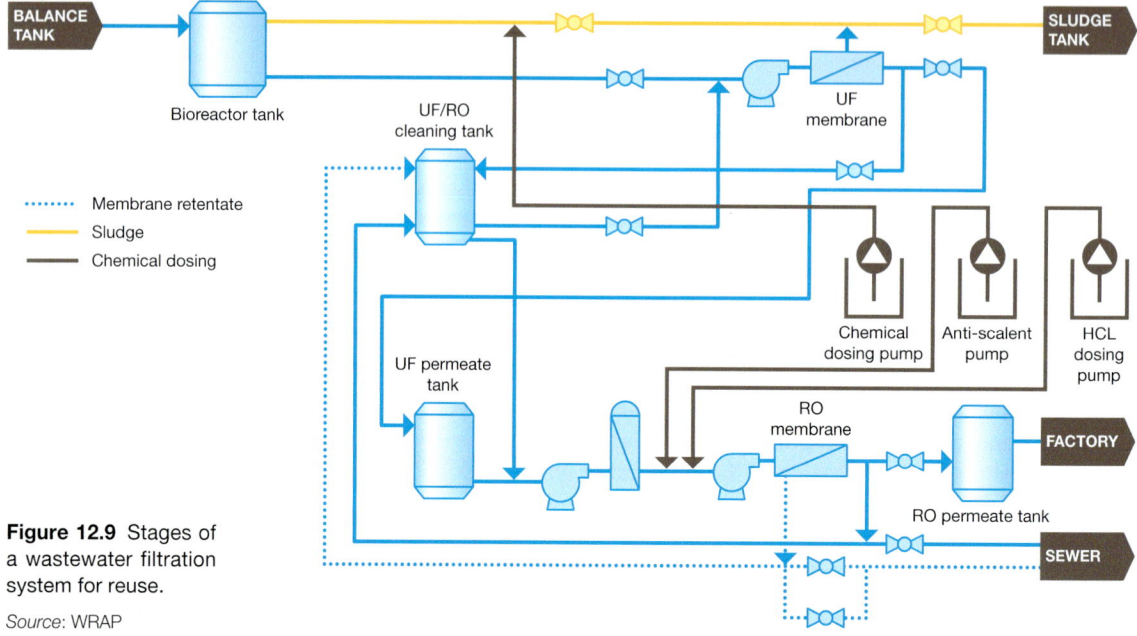

Figure 12.9 Stages of a wastewater filtration system for reuse.

Source: WRAP

Monitoring and analysis

In process industries where water is reclaimed and reused, it should be analysed to determine its contents, and it must be borne in mind that this can vary over time. The presence of chemicals, the water balance, the pH factor, conductivity, etc. may all have an unwanted effect when reused. Therefore particular forms of treatment may be required before it may be used and the quality of resultant treated water will need to be monitored, which all has a cost implication.

The reuse of secondary treated wastewater as cooling tower make-up using, for example, boiler blowdown is another reuse option that has been employed in many locations. Corrosion protection for the towers can be achieved when using a high-conductivity water source, but attention must be paid to mineral content. Conductivity in water is affected by the presence of inorganic dissolved solids such as chloride, nitrate, sulfate and phosphate anions (ions that carry a negative charge), or sodium, magnesium, calcium, iron and aluminum cations (ions that carry a positive charge). Organic compounds like oil, phenol, alcohol and sugar do not conduct electrical current very well and therefore have a low conductivity when in water. Conductivity is also affected by temperature: the warmer the water, the higher the conductivity. For this reason, conductivity is reported as conductivity at 25°C (77°F).

In-line demineralisers are used in such circumstances, and in many other industrial operations, to produce high-quality water for boiler feed water and other processes. The regeneration of demineralisers is usually accomplished in four steps: backwash, acid/caustic introduction, slow rinse, and fast rinse.

There are limitations to water reuse in an industrial environment. Since it involves taking water back for various uses somewhere in the process, this frequently involves evaporation, which means an additional concentration of salts. When used as cooling tower make-up or as scrubber water make-up, various alternatives are available to minimise the effect of these high-conductivity waters. These include chemical corrosion inhibitors, use of side stream softening, installation of different metallurgy, etc. As whole effluent toxicity testing takes place on many industrial dischargers, the impact of water reuse on toxicity must be included in any water reuse plan.

Many industrial facilities are mandated to use freshwater organisms, such as Daphnia magna, for effluent toxicity testing. Higher salinity levels in wastewater will cause osmotic stress to these organisms, so there is an unwanted effect between compounds or metals in the wastewater and the salinity level.

Wetland Energy Transduction systems

As an alternative to direct reuse, Wetland Energy Transduction (WET) systems are a form of wastewater treatment and recycling, using a variety of biological metabolic processes plus sunlight to create biomass. WET systems have been developed in the UK by Biologic Design since 1993. Over 150 systems have been created so far, varying in size up to one treating effluent for a population of 2,500. Industrial systems include cheese-making effluent, dairy farm wastewater and a very well established 8 acre system at Westons Cider in Herefordshire, England, which has been purifying very acidic wastewater for nearly 20 years. This has enabled the company to greatly expand its production capacity while saving millions of pounds that would have needed to be spent on conventional effluent treatment. Anyone viewing an established WET system (from about two years after planting) could imagine it to be a totally natural wetland that had been there for many years. They are designed to remain in place indefinitely.

Figure 12.10 Part of a freshly commissioned, nascent Wetland Energy Transduction system. Liquid effluents are processed naturally by microbes on the roots of selected wetland plants as they travel through a series of lined bunds.

Source: Author

Figure 12.11 A magnetic flow meter in use in a brewery.

Source: Wikimedia Commons (Mtaylor848)

Effluent monitoring

Effluent monitoring is required in other sectors too, such as the chemical, food and related industries. Timed analysis of the content of effluent related to the scheduling of production processes can identify where waste occurs. This implies the use of flowmeters, probes to measure pH, temperature, total organic carbon, suspended solids, COD (chemical oxygen demand) and so on. As with energy monitors, these can now be installed wirelessly. When pollution limits are approached, alarms can be activated. This can reduce both downtime and product losses.

Construction sites

Special advice on reducing water use is given for construction sites:

- For dust suppression a hydraulic spinning system can be 90 per cent more water efficient than a splash plate, provided mains-quality water is available. Chemical additives are an option to assist in reducing the volume of water needed.
- For demolition dust suppression, avoid high-capacity rain guns. Fan misting systems are a more efficient alternative.
- To reduce water used during building envelope and services commissioning and testing, build water recirculation and minimisation into the strategy. The water used for flushing building services should be isolated as soon as the flush water turns clear.
- Hoses must not be left running when not in use. Robust trigger guns should be fitted so that flow can be controlled at the point of use.
- For vehicle tyre washing, drive-through wheel washers can be chosen which recycle water in a closed loop. Waterless systems are an even better option; these use angled steel grids to clean debris from tyres.
- For washing out concrete wagons a high-pressure, low-volume efficient spray pattern can reduce water use. A specially designed sock can be fitted over the chute to reduce spills and eliminate pollution. Wash-out water may be reused at concrete batching plants.

Water-efficient products

The Water Technology List is a list of certified products in the UK that are among the most water efficient available. 'Enhanced Capital Allowances' may be claimed on their purchase. Visit www.eca-water.gov.uk. In America, the Watersense label fulfils a similar function. Products carrying the label are alleged to be at least 20 per cent more water efficient than average products in that category. They may be found on the United States Environmental Protection Agency website.

Rainwater harvesting

Rainwater may be collected, stored and used for toilets, washing machines, gardening and other purposes. It does not need disinfecting, merely filtering. The larger the roof, the more cost-effective the measure. In order to calculate the benefit, the annual rainfall for the location should be known, together with the roof area. This is then multiplied by a drainage factor which is dependent upon the roof type. The higher the roof factor, the greater the proportion of rain falling on the roof will reach the gutter and be collected. Since rainfall is sporadic, storage may be needed. The size of the tank should be sufficient to hold about 18 days' worth of demand, or 5 per cent of annual roof yield, whichever is the lowest.

Table 12.4 Approximate annual yield of rainwater in cubic metres per year for a range of roof sizes with varying rainfall

Plan roof area (m²)		50	75	100	125	150
Mm rain per year	500	15	22.5	30	37.5	45
	1000	30	45	60	75	90
	1500	45	67.5	90	112.5	135
	2000	60	90	120	150	180
Plan roof area (yd²)		59.8	89.71	119.60	149.5	179.40
Inches rain per year	20	17.94	26.91	35.88	44.85	53.82
	40	35.88	53.82	71.76	89.71	107.64
	60	53.82	80.73	107.64	134.55	161.46
	80	71.76	107.64	143.52	179.40	215.28

Table 12.5 Drainage factors for different roof types

Roof type	Drainage factor
Pitched roof tiles	0.75–0.9
Flat roof smooth tiles	0.5
Flat roof with gravel layer	0.4–0.5

Example

A roof has an area of 530m² (633.87 yd²). It receives 1,500mm (60 inches) of rain per year. The maximum collectable volume is therefore $90 \times 5.3 = 477m^3$ (624yd³). It is a flat roof so this is multiplied by 0.5, yielding a total of 238m³ (311.29yd³).

To calculate the tank size, multiply this figure by the filter efficiency (90%) and by 5 per cent. This results in 10.73m³(14yd³).

A system will involve the following:

- reliable guttering (steel or copper);
- downpipes;
- accessible filtration;
- frost-protected storage away from sunlight at a temperature which prevents bacterial growth;
- a floating intake to draw water from the top of the water surface so that sediment at the bottom is not collected (new water comes into the tank near the bottom);
- a clearly labelled separate system of pipes alongside the existing mains backup plumbing system to pump and direct the water to where it is going to be used within the house; the water should be oxygenated;
- a rat-proof overflow.

Figure 12.12 Three (5,000-litre or 22,730-gallon) rainwater harvesting tanks supplying the False Bay Ecology Park centre in South Africa with all its freshwater needs. From the tanks the water is pumped into the building with the pumps on the right.

Source: Cape Water Solutions

The pipework for this separate network of water supply should be clearly labelled throughout the system. Standards exist for this and should be followed.

Rainwater harvesting can be integrated with SUDS (sustainable urban drainage solutions). This means that, provided it is properly treated if pollutants are likely to be present, all surface water on a site can be potentially collected and reused. Surface water drainage systems would lead into a tank, from which the water can be pumped for reuse. The sizing of this tank would be calculated using a method that takes into account extreme weather conditions, such as once-every-50-year occurrences, or have provision for overflow into balancing ponds/swales. These would be open to visual inspection and periodic cleansing. Integrating the two systems, namely SUDS and rainwater collection, makes it a double win.

Greywater reuse

These systems reclaim water from places where it is not likely to be too contaminated, store it and usually reuse it in flushing toilets. Savings vary from 5 to 36 per cent. Treatment is necessary because warm, nutrient-rich greywater incubates bacteria when stored. This means that greywater should also only be stored temporarily. Electronically controlled dump valves can control this by emptying tanks after a period of inactivity before refilling them with mains water. Chemical disinfectants such as chlorine or bromine compounds may be added. In larger systems, the greywater may be treated in a small sewage treatment plant, using membrane filtration or ultraviolet technology. This is normally not cost-effective unless such a system already is in place.

Greywater systems require more frequent attention. Filter cleaning may be required if self-cleaning filters do not always function. They do have the advantage of a more predictable and constant supply of water than rainwater systems.

With both rainwater and greywater reuse, the carbon cost of creating and installing the system, plus running it using the pumps, should be taken into account before a decision is made. Any pumps used should be of the minimum power specification for the job and not oversized. If possible, gravity-fed systems

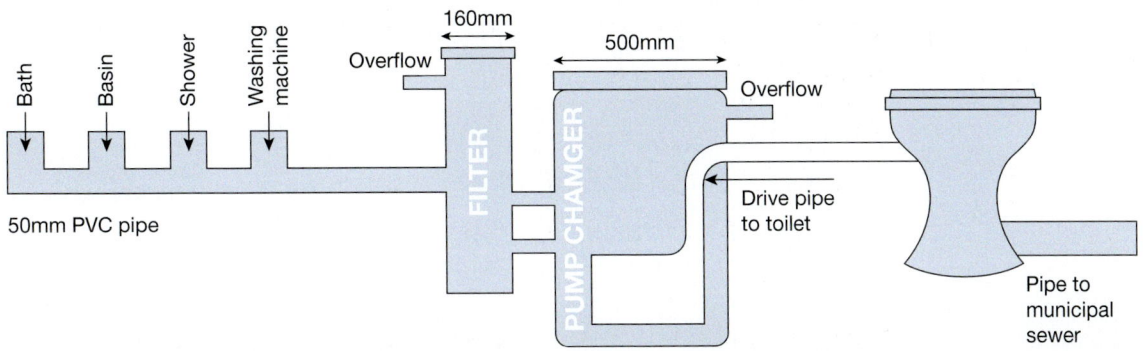

Figure 12.13 Basic components of a greywater reuse system.

Source: Cape Water Solutions

should be deployed to minimise or remove the need for a pump. In both cases as well, using the water for irrigation purposes is the simplest end use.

Notes

1 Carbon Trust survey by telephone with 475 C-level executives from a variety of functions within a wide-ranging number of industries. Interviews were conducted by Vanson Bourne during October 2012.
2 Water Key Performance Indicators and benchmarks for offices and hotels. C657 CIRIA (http://www.ciria.org).
3 Some buildings will contain a mix of uses (e.g. office/staff areas and public spaces such as retail floors). In this instance, benchmarking should be based on the dominant use type (if clear) or independently for each type of space if each use is significant.

13
Making the financial case

Among the barriers often found to the implementation of energy management measures, persuading senior management to make investments and commit to projects is one of them. A lack of understanding, conflicting priorities, misaligned financial incentives, hassle cost and lack of financial backing figure in this situation. This chapter looks at several ways of presenting and comparing the financial case for different measures in order to overcome these barriers.

Marginal abatement cost curves

A marginal abatement cost curve (MACC) is a helpful visual aid to providing an idea of the annual potential to reduce emissions and the average costs of doing so for a wide variety of technologies. We first met them in the Introduction. MACCs are a useful tool for cost-effectiveness analysis. But how are they compiled?

In Britain, the Committee on Climate Change (CCC)[1] has produced several MACCs for energy efficiency that incorporate research including that generated by ENUSIM (the Industrial Energy End Use Simulation Model), originally designed to model industrial energy use by considering the take-up of energy-saving technologies in industry.

The MACC for the non-domestic sector is illustrated in Figure 13.1. The CCC concludes that, for the UK as a whole, there is 'a very significant contribution from improved energy management. Many measures are almost entirely low cost, with the potential to save over £800m countrywide per year for firms with very little (if any) up front expenditure. They could save over 8 $MtCO_2$ per year.'

Estimating payback

MACCs are arrived at by calculating the payback for various measures. Projects are usually sold to management on the basis of return on investment. This may be expressed in two ways: (1) as an effective interest rate, based on the net present value; and (2) as a payback period, i.e. the length of time it takes for the initial investment to be recouped by the savings earned or income generated.

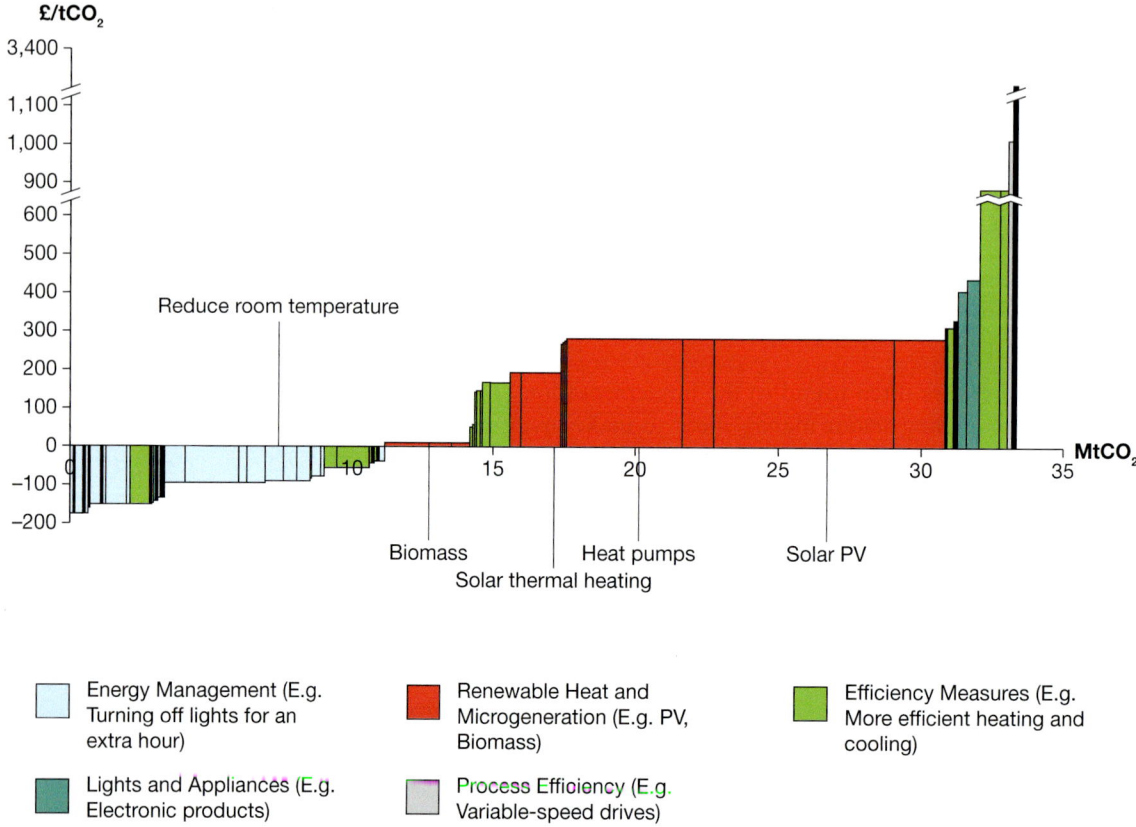

Figure 13.1 A marginal abatement cost curve (MACC) illustrating the technical potential for improvements in the non-domestic sector. Each column represents a particular measure. The vertical axis represents the cost per ton of carbon dioxide saved. The horizontal axis represents megatons of carbon dioxide saved throughout the lifetime of the measure. Measures to be taken on the left of the graph with columns descending beneath the horizontal axis have a negative cost; i.e. they save money. The ones on the right with columns ascending above the horizontal axis have a net cost; i.e. they cost more than they save. The further right that a measure is positioned, the greater its lifetime cost. All energy management measures have a negative cost and save money, as do many efficient heating and cooling methods.

Source: Committee on Climate Change

Simple payback

The most basic of these is simple payback. However, it does not always illustrate the true benefits of an investment. Suppose an organisation demands a two-year payback period from any investment. Then, as the following example shows, it would miss out on the benefits of a project with a six-year payback period that actually had a better return on investment.

A project costing £60,000 which receives £30,000 in benefit per year following completion but which only lasts for three years would yield a total of £90,000. A project which costs the same amount, but only yields £22,000 per year, yet lasts for six years, would yield a total of £132,000. However, if it were only evaluated on a two-year basis it would lose out to the three-year project.

A project which repays its cost every three years is demonstrably better than one which promises to return the investment in three years. To help establish this, the concept of discounted cash flow is introduced.

Discounted cash flow (DCF)

Discounted cash flow provides a more realistic way of establishing payback. There are three stages to estimating DCF:

1   estimate the resulting cash flow;
2   apply the discount rate;
3   calculate the end value (net present value).

The cash flow is taken from the estimated savings in energy cost resulting from the measure taken. This will depend upon projections of future energy cost. For

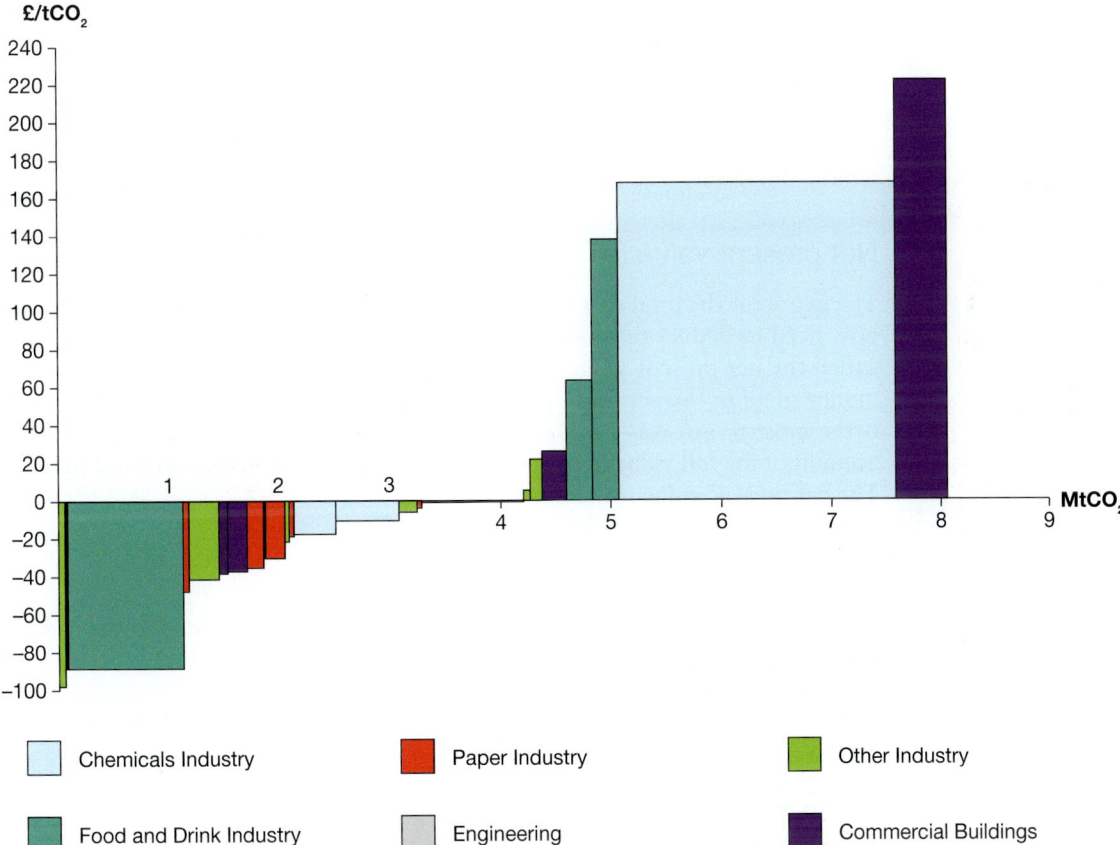

Figure 13.2 A marginal abatement cost curve (MACC) illustrating the potential for CHP (combined heat and power) in different sectors. It shows that, even within a sector, whether a particular project is cost-effective depends on individual conditions. This is why, for each sector, there are different instances (illustrated by columns of the same colour), some of which are above the line (net cost) and some below the line (net benefit).

Source: CCC

example, energy prices over the past three years can be projected on a median basis into the future, but this will then need to be discounted at a chosen discount rate. Discount rates are a function of the rate of inflation and represent what one unit of currency will be worth in a year's or ten years' time. An average price [P] is calculated this way for each year of the projected lifetime [L] of the project. Each of these figures is then multiplied by the amount of energy [E] expected to be saved every year.

The lifetime period chosen for the project will depend upon the expected lifetime of the technology. If it were a boiler, for example, it could be 15 years. Should it be an insulation measure, it could be 30 years. The total cost savings [S] generated by energy not used compared to not doing the project, over the lifetime of the project, will then be:

$$S = E \times [P_{(year\ 1)}] + E \times [P_{(year\ 2)}] + E \times [P_{(year\ 3)}] \ldots E \times [P_{(year\ L)}]$$

What discount rate should be chosen? The industrial model ENUSIM uses private fuel prices and a 10 per cent discount rate to reflect the incentives faced by firms. Some UK organisations adopt the rate used in the UK government Treasury's Green Book, which sets out the framework for the evaluation of all policies and projects, namely 3.5 per cent. Others simply adopt the current rate of inflation, or interest rate on a loan taken out for the purpose of the measure that would need to be repaid. It is useful to run the calculation several times with different discount rates.

Net present value (NPV)

The figure for the total cost savings [S] is not the final step in our calculation. We now need to deduct the cost [C] of taking the measure, which gives us a figure called the net present value [NPV] of the project. This is the value in today's money of all of the net profit that will be generated from taking this measure. It is the most useful way of comparing the value of different measures. It takes account of the full value of the project and presents it in easily comparable form. The net present value is therefore:

$$NPV = S - C$$

This is how all of the figures were arrived at that are represented in the MACC graphs above. Applying this to the two projects above, with a 10 per cent discount rate, allows us to see the following:

Project 1 yields:

£30,000 (year 1) + £27,000 (year 2) + £24,300 (year 3) = £81,300, not £90,000

Project 2 yields:

£22,000 (year 1) + £19,800 (year 2) + £17,820 (year 3) + £16,038 (year 4) + 14,434.20 (year 5) + £12,990.78 (year 6) = £103,082.98, not £132,000

Both projects cost the same: £60,000. Subtracting this from the cost savings reveals that the NPV of the first is just £21,300, while that of the second is £43,082.98, i.e. over double.

Internal rate of return (IRR)

The NPV will also allow the projects to be compared to what would happen to the same amount of money were it to be invested in a bank account with the same interest rate as the discount rate chosen. This is done by calculating the internal rate of return (IRR), or the interest rate on the investment, and is easily accomplished using Microsoft Excel as follows (and see Figure 13.3):

1   The initial expenditure is typed into a cell on a spreadsheet. This must be a negative number. Using our original example, –60,000 would be typed into the A1 cell.
2   The subsequent discounted cash return figures above for each year are entered into the cells directly under the first cell. Following the example in Project 1, this would mean typing 30,000 into cell A2, 27,000 into cell A3, etc.
3   The IRR is then revealed by typing into the next cell beneath all the values the function command '=IRR(A1:A4)' and pressing the enter key. In this case, the IRR value, 18 per cent, is then displayed in that cell.

The IRR of the second project, calculated by the same method, is 20 per cent, and so provides a better rate of return. It is relatively easy to set up a template in Microsoft Excel to enable the performance of a similar calculation for any capital investment project. Further costs that are unique in any given year may be added, such as figures for additional maintenance, additions or repairs, and, at the end of the project, a figure for resale of any equipment, for example, its scrap value.

Presenting projects in such a way to senior management will allow them to compare their value with other projects they may be considering, as well as enabling the energy manager her- or himself to prioritise projects.

Ways of offsetting risk

It is still possible that management will argue that the initial capital outlay for the project cannot be justified, despite long-term benefits. In this case there are other possibilities. First, it is worthwhile considering asset finance such as leasing and renting. These techniques offset the monthly cost of the new equipment against the energy savings it delivers across the financing term, effectively making it a zero net cost or even a cash-positive investment.

	A	B	C	D	
1	-60000				
2	30000				
3	27000				
4	24300				
5	18%				

A5   fx =IRR(A1:A4)

Figure 13.3 Using Microsoft Excel to calculate the internal rate of return of an investment. The formula in the field at the top is entered into cell A5 and yields the percentage rate based on the figures above.

Source: Author

Second, it may be possible to obtain the services of an Energy Services Company (ESCo), which would bear all the risk. This company then sells the service to the client under a long-term contract. The service could be heat, lighting, power, or a whole package which includes efficiency and energy management. In the USA, these companies offer energy savings performance contracts, under which they develop, install and arrange financing for improvements to boost energy efficiency and lower costs.

Third, land or roof space may be leased to a local utility or energy firm. They install the equipment, for example, solar panels or a CHP plant, and sell the electricity, reaping the benefit of any tax credits or subsidies. They then pay rent to the client, and possibly sell a proportion of the energy generated at a discount. In the UK these arrangements come under the heading of power purchase agreements.

Finally, grants, tax credits and subsidies may be available at a local, national or federal level. In the USA, the Department of Energy's Energy Efficiency and Renewable Energy Program coordinates funding for projects. There are several tax credits, grants and rebates available. For example, a tax credit scheme is on offer until 2016 which gives 30 per cent credit on biomass, heat pumps, solar, wind and fuel cells. Energy-efficient mortgages are available also from the government and some private loan companies, which help buyers to renovate an inefficient property or build a new one. Different states offer different incentives, which vary widely. For instance, New York state offers grants of up to 50 per cent for renewable microgeneration. A good source of information is the Database of State Incentives for Renewables and Efficiency (DSIRE, at www. dsireusa.org). In some states there are electricity buy-back credits available for renewable energy fed back into the grid. Information is available from the local utility.

In the UK, the Carbon Trust is a good source of information on tax credits and other incentives. Feed-in tariffs, Renewable Obligation Certificates and the Renewable Heat Incentive are other schemes. Some projects may be eligible for finance from the Green Investment Bank at favourable rates, while specialist banks such as Triodos and the Cooperative Bank look especially favourably upon schemes with environmental aspects.

Measurement and verification (M&V)

Measurement and verification (M&V) (see Chapters 1 and 2) demonstrates how much energy the measure has avoided using, a prerequisite for evaluating the total cost saved. There are four possible ways of doing this:

1　With respect to actual measurements of a particular performance indicator, such as the power used by a particular appliance, set of appliances, lighting, etc. that are being replaced, plus estimates based, say, on models or historical data, and all other variables.
2　Based on the actual measurement of all performance indicators.
3　Based on the actual measurement of energy use of a whole facility, or sub-facility that is affected by the measure.
4　Based on modelling or simulation of the energy use of the facility or sub-facility; this is employed where there are no historical data.

Evaluations must be as accurate as possible; for example, there may be additional electrical costs associated with the use of heating equipment. Transport fuel costs may need to be taken into account. Care should be taken to make sure that reporting is not over-optimistic. Transparency should be the watchword.

The International Performance Measurement and Verification Protocol (IPMVP), managed by the Efficiency Valuation Organisation, an international non-profit organisation dedicated to supporting M&V, is the globally accepted standard for quantifying the results of energy efficiency investments, demand management and renewable energy projects. It was created as an initiative of the United States Department of Energy. IPMVP is used to measure actual results against those modelled, as well as benchmarks.

ISO 50001, the internationally accepted energy management system standard, requires that an M&V plan be put in place and followed for any measure. In addition, the Efficiency Valuation Organisation runs accredited training in many countries in association with the Association of Energy Engineers leading to the qualification Certified Measurement and Verification Professional (CMVP). This means that third-party independent evaluation of a project can be made if required, for example, for reporting purposes (CSR and annual reports).

Whether there is independent evaluation or not, the presence of a plan helps to preserve confidence in the investment that is to be made. If an energy manager wishes to remain successful and to build their reputation within the organisation, and, indeed of the organisation, they must build up trust, and this is one way to do so. The promised savings must actually be either delivered or exceeded. That is the bottom line.

It follows that the cost of M&V needs to be built into the cost of the project. If, for example, a BEMS is already installed, it forms part of the overheads. If the project includes the installation of a BEMS, then part of the payback will be the benefit of being able to monitor further projects.

Carbon strategy

Most companies now have a carbon-saving strategy with targets for reduction alongside financial targets. They need to record and measure their carbon emissions, and predict outcomes. Most of the BEMS and software described in Chapter 2 will have the facility to convert energy use into carbon emissions, although data will have to be brought in from elsewhere as well, for example, transport. The energy manager's case to management must therefore include quantified energy savings and projections, just as it does financial savings. The carbon emissions abated by various measures varies according to the mix of supply in the local grid at that particular location.

Some organisations present CO_2 emissions in tonnes of carbon instead of tonnes of CO_2. To convert from tonnes of carbon, multiply by 44/12, which is the molecular weight ratio of CO_2 to C.

Political strategy

Like any senior manager, an energy manager needs to be a consummate strategist. He or she is competing for funds for projects with other managers in other departments. There are several tactics that may be used:

- cultivating an ally on the board;
- building up trust by first proposing no-cost or low-cost projects and quick wins;
- not proposing any project which has not been thoroughly investigated and costed, and discussed with the people who would be affected by it;
- looking at which departments or sites have the most potential for energy savings;
- having several projects available at any one time to propose, with different advantages;
- being able to demonstrate or conduct visits to successful similar projects elsewhere;
- being able to demonstrate a competitive advantage to be gained from the project. This could be qualitative as well as quantitative, such as improvement in working conditions or an improved reputation for the organisation;
- allying the project with the aims of the organisation as a whole, for example, the attainment of corporate targets. This might include setting an overall publicly declared energy target for the organisation.

The value of site visits

Financial reasons are not the only potential barrier to take up of low carbon technologies. One of the best ways found in the extensive research on this subject to persuade people to take action is if they see a particular measure successfully employed elsewhere. Energy managers may find it beneficial to organise trips to such sites as exemplars of measures they have in mind, so that the reality may be brought home most dramatically. The testimonies and recommendations of those who have already taken these steps, and the ability of visitors to question them to allay any doubts and dispel any misconceptions, are invaluable.

Notes

1   Mark Weiner, *Energy Use in Buildings and Industry: Technical Appendix*, Committee on Climate Change, London, March 2009.

14
Conclusion

The world is now firmly in the age of resource efficiency, of which energy efficiency forms a part. For an organisation to be able to survive into the future, it has therefore to see all of its operations – its requirements in terms of materials, energy and water, its fixed assets – as equal in importance to its core activity. Unilever is one company that is leading the way. Its CEO, Paul Polman, is the visionary behind a Sustainable Living Plan, launched in November 2010, which seeks to double sales and halve the environmental impact of its products. It is working. He believes that this fundamental shift in the business paradigm is partly a reaction to the financial crisis, from a rules-based one back to a principles-based one, but it has financial benefits. It follows that a strategy like this should translate into a procurement strategy.

Procurement of new equipment

It should be part of any procurement strategy to purchase equipment that is sustainable and consumes the least energy, or has the least environmental impact, over its lifetime compared to compatible products. Lists of these products, together with standards, may be found on the website of the U.S. Energy and Efficiency and Renewable Energy office, or the European Market Transformation Programme website (http://bit.ly/18uIOEg), with further information on the Energy Using Products Directive website (www.eup-network.de).

Standby power load should also be a choice factor. For instance, according to this office, US federal agencies must purchase products with a standby power level of 1W or less. Standby power typically occurs when the product is switched off for not performing its primary purpose. The standby power data centre (www.1.usa.gov/hVxDmC) lists compliant products.

Sustainable procurement is a specialism in itself. Specimen framework agreements to ensure the supply of sustainable goods and services are available from the website of the UK Sustainable Procurement Centre of Excellence. There, you will also find a knowledge base of information on sustainable procurement, commodity areas, carbon reduction, whole life costing, legislation, toolkits, case studies and best practice.

For ICT, ENERGY STAR® is a voluntary labelling scheme for products which use less than a specified energy consumption in typical use. It was originally developed by the US Environmental Protection Agency (EPA) for common computing equipment, but is now a joint activity with the European Commission.

This means that specification of ENERGY STAR compliance in tenders is compatible with EU procurement rules. A list of compliant models is at www.eu-energystar.org.

A key factor in the energy consumption of ICT is user behaviour. It is now possible to purchase software which monitors actual PC use within an enterprise. Advanced versions permit energy managers to set power policies that reflect a certain level of usage, reducing both energy consumption and carbon emissions. They reveal when users are using their PCs by monitoring key strokes and mouse movements. Energy managers can then match the power state of each subset of PCs, by location, with the activity level of employees. They can also identify unused or underutilised PCs on the network, further eliminating the management overheads of maintaining these machines; ensure that a computer in a low-power state can be woken up and accessed on demand when a user is working remotely; and that applications which prevent a PC from being powered down can be overridden while the PC is not in use.

Supply chain optimisation

It then becomes necessary for businesses and organisations seeking to reduce their carbon footprint to turn attention to that of their products and services, and this involves looking at their supply chains. According to the American Council for an Energy-Efficient Economy (ACEEE), supply chain optimisation can result in up to 60 per cent of energy intensity reductions.[1] For example, in food production and distribution, much perfectly good food is wasted due to spoilage, both in the supply chain and at the retail level. This means that all of the energy embedded in the food is wasted as well. By modelling the supply system throughout the chain, opportunities may be identified to significantly reduce waste by changing processing, handling, packaging and delivery systems. The result is frequently fresher food delivered faster and of a more consistent quality. There is less waste and greater savings.

Case study: PepsiCo, UK

Snack foods manufacturer Walkers, and its parent company PepsiCo, have been working with the Carbon Trust on energy efficiency and carbon management. They have saved over 2,000 tonnes of CO_2 per year, reducing energy bills by approximately £225,000 (US$350,000). Having done this they moved on to looking at their supply chain in order to demonstrate a continuing commitment to emissions reduction. They began by looking at their raw material production, which includes potato and corn producers, sunflower oil and vegetable oil manufacturers, corrugated cardboard manufacturers and so on. They then began to optimise the distribution of raw materials using logistics and network planning. They have already optimised the manufacture of products. The next step was product distribution, again tackled by the network strategic planning department. Finally, they wanted to make sure that redundant packaging could be recycled.

Resource efficiency and materials substitution

Most industries transform materials from one form to another. The shift from using virgin feedstocks to recycling or reusing existing materials can result in significant energy savings. Moreover, as many feedstocks and raw materials become increasingly scarce, savings will be found, especially in countries like Britain where sending materials to landfill is taxed. Plastics recycling is a case in point.

Other examples, cited by ACEEE, include the use of rubberised asphalt, which lasts twice as long and requires half the volume of conventional asphalt, the use of pozzolans (cement extenders) and non-Portland cements to displace Portland cement in structural concrete, and the use of waste products from manufacturing processes as feedstock for other products. High-strength glass fibres enable the manufacture of lightweight composites at low cost relative to carbon fibre.

The use of waste products from other processes, industries or companies is called 'industrial symbiosis'. As in nature, where one organism will thrive on the wastes of another organism, and there is no overall waste inside an ecosystem, industrial ecosystems are moving in the same direction to create zero waste closed-loop economies. In many geographical areas there may be existing networks, or there may be possibilities for setting them up where they don't already exist, under which members can offer waste products that are otherwise costly to dispose of, and sell them at profit to other members.

Case study: Low carbon tomatoes

Figure 14.1 Terra Nitrogen's tomato greenhouse.

Source: Gareth Kane

Terra Nitrogen, a company based in Billingham in the northeast of England, which produces nitrogen chemicals and methanol for industry, as an unfortunate by-product also produced a lot of carbon dioxide emissions. It linked up with John Bader Ltd, which now diverts carbon dioxide from the plant into 38 acres of greenhouses erected next door to grow tomatoes. Terra Nitrogen is also supplying electricity to the greenhouses, allowing them to continue production through the winter and removing the need for the UK's supermarkets to import so many tomatoes from Spain. The benefits include the successful reuse of waste heat, reduction of 12,500 tonnes of carbon dioxide emissions and the creation of 65 new jobs.

Disruptive developments

Prospects are emerging for new processes to transform what is manufactured and the methods by which this is done. In the iron and steel industry, direct iron production, whereby iron is produced from iron ore using a reducing gas without the need to produce coke, reduces the energy and carbon emissions resulting from the latter process. Organic chemicals and industrial gases are also nowadays being produced in less energy-intensive ways. Nanotechnology is offering energy-saving opportunities. In metal and glass manufacturing, submerged combustion melting processes can reduce fuel use by 20 per cent.

Watching out for such disruptive, game-changing developments is not really the job of most energy managers. Nevertheless, it does no harm to keep a watchful eye for such opportunities. The most effective energy manager in a particular industry knows that industry very well and is in a good position to advise management on strategies for adaptation in this fast-changing world.

Technology is only part of the solution. The rest is down to people and organisational culture. Engaging the workforce and management in the campaign to reduce energy consumption has very little capital cost but achieves great benefit. The most obvious energy reduction opportunities are often the most overlooked. Energy consultant Kit Oung lists some of these as follows:

- lighting;
- excessive ventilation compared to demand;
- simultaneous heating and cooling;
- idle time in production;
- generation of graded or waste products;
- poorly executed maintenance resulting in subsequent re-failure.

Oung notes that there are other benefits from implementing such measures. For example, turning down the air supply can make the workplace quieter. Improving lighting, in particular by adding daylighting, makes for a more pleasant working environment. Reducing the energy consumption of machinery can make it last longer. The energy manager can therefore see a way to play their part by not only reducing environmental impact and saving money, but improving working conditions for everyone affected.

Note

1   Anna Shipley and R. Neal Elliott, *Ripe for the Picking*, 2006, Washington, DC: American Council for an Energy-Efficient Economy.

Appendix

Energy use

Units

kilo-	k	10^3	10,000
mega-	M	10^6	10,000,000
giga-	G	10^9	10,000,000,000
tera-	T	10^{12}	10,000,000,000,000
peta-	P	10^{15}	10,000,000,000,000,000

For example:

milliwatt (mW): 1000th of a watt
kilowatt (kW): 1,000W
megawatt (MW): 1,000,000W
gigawatt (GW): 1,000,000,000W
terawatt (TW): 1,000,000,000,000W. In 2006 about 16TW of power was used
worldwide.

Abbreviations

Btu	=	British thermal unit (MBtu = millions of Btus)
MJ	=	megajoule
TJ	=	terajoule
Gwh	=	gigawatt-hours
toe	=	tonnes of equivalent oil (Mtoe = millions of toe)
Kcal	=	kilo calorie
Gcal	=	giga calorie

Conversion factors

To	TJ	Gcal	Mtoe	MBtu	GWh
From	**Multiply by:**				
terajoule (TJ)	1	238.8	2.388×10^{-5}	947.8	0.2778
gigacalorie (Gcal)	4.1868×10^{-3}	1	10^{-7}	3.968	1.163×10^{-3}
million tonne of oil equivalent (Mtoe)	4.1868×10^{4}	10^{7}	1	3.968×10^{7}	11,630
million British thermal unit (MBtu)	1.0551×10^{-3}	0.252	2.52×10^{-8}	1	2.931×10^{-4}
gigawatt-hour (GWh)	3.6	860	8.6×10^{-5}	3,412	1

From	to kWh. Multiply by:
Therms	29.31
Btu	2.931×10^{-4}
MJ	0.2778
Toe	1.163×10^{4}
Kcal	1.163×10^{-3}

For example:

Conversion of 100,000 Btu to kWh:
100,000 Btu = 100,000 $\times$ 2.931 $\times$ 10-4 kWh = 29.31kWh

Conversion factors for mass

To:	kg	T	lt	st	lb
From	**Multiply by:**				
kilogram (kg)	1	0.001	9.84×10^{-4}	1.102×10^{-3}	2.2046
tonne (t)	1,000	1	0.984	1.1023	2,204.6
long ton (lt)	1,016	1.016	1	1.120	2,240.0
short ton (st)	907.2	0.9072	0.893	1	2,000.0
pound (lb)	0.454	4.54×10^{-4}	4.46×10^{-4}	5.0×10^{-4}	1

Conversion factors for volume

To:	gal U.S.	gal U.K.	bbl	ft3	l	m³
From:	Multiply by:					
US gallon (gal)	1	0.8327	0.02381	0.1337	3.785	0.0038
UK gallon (gal)	1.201	1	0.02859	0.1605	4.546	0.0045
barrel (bbl)	42.0	34.97	1	5.615	159.0	0.159
cubic foot (ft³)	7.48	6.229	0.1781	1	28.3	0.0283
litre (l)	0.2642	0.220	0.0063	0.0353	1	0.001
cubic metre (m³)	264.2	220.0	6.289	35.3147	1,000.0	1

Carbon dioxide emission factors by gross calorific value

Energy source	KgCO$_2$/kWh	KgCO$_2$ per other units
Natural gas	0.18523	5.3808 per therm
LPG	0.21445	6.2915 per therm
Coal	0.32227	2,383 per tonne
Diesel	0.25301	3,188 per tonne
Petrol	0.24176	2.3117 per litre
Fuel oil	0.26592	3,228 per tonne
Burning oil	0.24683	3,165 per tonne
Wood pellets	0.03895	1,83.9 per tonne

The greenhouse gas conversion factor comprises the effect of the CO2, CH4 and N2O combined, quoted as kgCO2e per unit of fuel consumed.

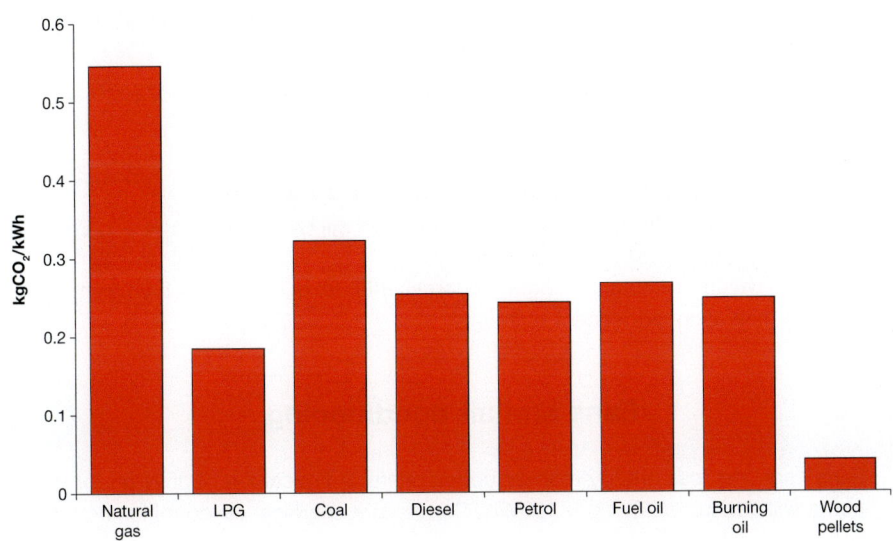

Figure 15.1
Comparison of the global warming potential of various fuels, showing kilograms of CO$_2$ produced per kWh of energy generated.

Source: Author

The global warming value for electricity varies by time, country and location, depending on the fuel mix. Figures are available from the Independent Energy Agency at www.iea.org/publications/freepublications/publication/name,4010,en. html.

Power and energy

Power is the rate at which energy is produced by a generator or consumed by an appliance.

> *Unit*: the watt (W). 1,000 watts is a kilowatt (kW).

Energy is the amount of power produced by a generator or consumed by an appliance over a period of time.

> *Unit*: the watt-hour (Wh). 1,000 watt-hours is a kilowatt-hour (kWh), commonly a unit of electricity on a bill.

> *Alternate unit*: the joule (J). Watt-hours may be used to describe heat energy as well as electrical energy, but joules are also used for heat. 3,600 Joules = 1Wh. Put another way, a joule is one watt per second, since there are 3,600 seconds in an hour; or 3.6 megajoules (MJ) = 1kWh.

Examples:

- One 80W light bulb on for two hours, or two 80W bulbs on for one hour would consume 2 × 80 = 160Wh.
- Three 80W light bulbs on for six hours will consume 3 × 80 × 6 = 1,440Wh or 1.44kWh.

Energy efficiency and lighting

For the same amount of luminescence over four hours:

- An 80W incandescent bulb will consume 80 × 4 = 320Wh.
- A low-energy 18W compact fluorescent bulb will consume 18 × 4 = 72Wh.

From this we can see that a compact fluorescent bulb is 72/320 = 4.5 times more efficient than an incandescent bulb.

Fans and air conditioning

Fans consume less energy than air conditioners. A typical fan of 30W on for six hours will consume 30 × 6 = 180Wh. Conversely, a typical air-conditioning unit of 2kW or 2,000W on for the same period will consume 2,000 × 6 = 12,000Wh. That is 67 times more energy.

Power generation

- One photovoltaic solar panel producing 80W for two hours, or two panels producing 80W for one hour would produce 2 × 80 = 160Wh.
- Three panels producing 90W for five hours will produce 3 × 90 × 5 = 1,350Wh or 1.35kWh.

Insulation

There is a relationship between the thermal conductivity of any material, its thermal resistance (the R-value) and its heat transfer (the U-value) properties, which all relate to the standard of insulation we want for a low carbon building.

Thermal conductivity (k)

Thermal conductivity, k (also known as psi or denoted λ), tells us how well a material conducts heat. The figures are supplied by manufacturers. It is:

$$k = Q/T \text{ times } 1/A \text{ times } x/T$$

or the quantity of heat, Q, transmitted over time t through a thickness x, in a direction perpendicular to a surface of area A, due to a temperature difference, T. The units used are either SI: W/mK or in the US: Btu/(hr × ft × °F). To convert, use the formula 1.730735 Btu/hr × ft × °F = 1 W/mK.

R-value

The R-value is a measure of how well a material resists heat travelling through it. It is the ratio of the temperature difference across an insulator and the heat flow per unit area through it. The bigger the number the better the insulator. It is the depth/thickness of a material divided by its thermal conductance; in other words, R = d/k.

To compare two insulants with different thicknesses and thermal conductivity, it is necessary to calculate the value of R for each.

R-values are given in metric units: square-metre Kelvin per watt or $m^2 \times K/W$ (or equivalent to $m^2 \times °C/W$); or, in the United States, in $ft^2 \times °F \times h/Btu$. It is easy to confuse them because R-values are frequently cited without units, e.g. R-3.5. One R-value (US) is equivalent to 0.1761 R-value (metric), or one R-value (metric) is equivalent to 5.67446 R-value (US). Usually, the appropriate units can be inferred from the context and their magnitudes.

Doubling the thickness of an insulating layer doubles its thermal resistance. R-values are often used when there are multiple materials through which heat will travel. The R-values of adjacent materials can be added together to calculate the overall value; e.g. R-value (brick) + R-value (insulation) + R-value (plasterboard) = R value (total). Another way of calculating it is to add the inverse of the k values of each element multiplied by their thickness, or: R(total) = (1/k) × d. Remember to include internal and external resistances and unvented air gaps.

U-value

R-value is the reciprocal of U-value (and vice versa of course). A lower U-value indicates greater insulation value. Commonly used in Europe, it is the overall heat transfer coefficient, describing the rate of heat transfer through a building element over a given area, under standardised conditions. The usual standard is at a temperature gradient of 24°C, at 50 per cent humidity with no wind.

It is described in watts per square metre Kelvin (W/m^2K or the amount of energy lost in watts per square metre of material for a given temperature difference of 1°C or 1°K from one side of the material to the other. Another way of understanding it is to see it as thermal conductivity divided by the depth of insulation, or U = k/d where k is the thermal conductivity of a material, d the material's depth.

Building regulations provide minimum standards of thermal insulation, typically expressed as a U-value for a given building element. This is found by adding the U-values for the different materials times the depth and area used for each within the element. In each case, measurements are taken on-site and then reference is made to information tables for the purpose of the calculation.

Calculating the R-value or U-value of an entire building

To calculate the R-value of a complete wall, it's necessary to add the U-values of each section (e.g. parts containing studs and those parts without studs, lintels, etc.) multiplied by the percentage of the overall wall they represent, and then inverse the result. This process is repeated for each building element (walls, roof, floors). These complex calculations are undertaken with bespoke software such as the free one at www.thermalcalconline.com.

Calculating heat losses and gains

There are two primary methods of heat loss or gain in building: conduction through the building envelope; and air movement through and between the elements.

The general heat loss/gain formula is: Q = U*A*ΔT, where the heat loss of an area of size A is determined by the U-value of the materials and the difference in temperature between inside and out (that is the difference in temperature of the two surfaces, not the two air temperatures).

Index